AUSTRALIAN WRITERS

PETER PORTER

The *Australian Writers* series examines in succinct form the lives and works of Australia's major writers. This authoritative series, under the General Editorship of Chris Wallace-Crabbe, will become essential reading for all who are seriously interested in Australian literature. *Australian Writers* is a timely and expansive successor to Oxford's earlier *Australian Writers and Their Works* series, and takes account of new cultural formations and developments in literary theory. At the same time, these provoking books contain elements of biography and are aimed at the intelligent general reader.

General Editor

Chris Wallace-Crabbe is the holder of a Personal Chair at the University of Melbourne. Since 1989 he has been Director of the Australian Centre. Educated at Melbourne and Yale Universities, he has an international reputation as a poet. Recent books include *For Crying out Loud*, a collection of poems, and *Falling into Language*, a collection of critical essays. In 1987–88 he held the Visiting Chair of Australian Studies at Harvard.

OXFORD

AUSTRALIAN WRITERS

PETER PORTER

PETER STEELE

Melbourne

OXFORD UNIVERSITY PRESS

Oxford Auckland New York

OXFORD UNIVERSITY PRESS AUSTRALIA

Oxford New York Toronto
Delhi Bombay Calcutta Madras Karachi
Kuala Lumpur Singapore Hong Kong Tokyo
Nairobi Dar es Salaam Cape Town
Melbourne Auckland
and associated companies in
Berlin Ibadan

OXFORD is a trade mark of Oxford University Press

First published 1992

National Library of Australia
Cataloguing-in-Publication data:

Steele, Peter, 1939– .
Peter Porter.

ISBN 0 19 553282 1.
ISBN 0 19 553288 0 (series).

1. Porter, Peter, 1929– . 2. Porter, Peter, 1929– —Criticism and interpretation. 3. Poets, Australian—20th century—Biography. I. Title. (Series: Oxford Australian writers series).

A821.3

Typeset by Solo Typesetting, South Australia
Printed by Impact Printing Victoria Pty Ltd
Published by Oxford University Press,
253 Normanby Road, South Melbourne, Australia
Cover photograph courtesy of Jill Neville

CONTENTS

A NOTE ON TEXTS AND ABBREVIATIONS

Page references for quotations are taken from the editions listed in the Bibliography. Quotations of poems which appeared in Peter Porter's *Collected Poems* (1983) refer to this source. The page reference in the *Collected Poems* is preceded by the volume in which the poem first appeared. Quotations from poems in volumes published after 1983 refer to those volumes alone.

The following abbreviations have been used for volumes of poetry by Peter Porter:

CP	*Collected Poems* (1983)
OBTB	*Once Bitten, Twice Bitten* (1961)
1962/3	*Poems in Penguin Modern Poets 2* (1962) and *A Group Anthology* (1963)
PAAM	*Poems Ancient and Modern* (1964)
APF	*A Porter Folio* (1969)
TLOE	*The Last of England* (1970)
PTTC	*Preaching to the Converted* (1972)
AM	*After Martial* (1972)
LIACC	*Living in a Calm Country* (1975)
TCOS	*The Cost of Seriousness* (1978)
ES	*English Subtitles* (1981)
FF	*Fast Forward* (1984)
TAO	*The Automatic Oracle* (1987)
PW	*Possible Worlds* (1989)
APS	*A Porter Selected* (1989)
TCOB	*The Chair of Babel* (1992)

CHRONOLOGY

1929	Peter Porter born in Brisbane, only child.
1934	Begins school, Junction Park State School.
1938	Death of his mother.
1939–40	Church of England Grammar School.
1941–46	Attends Toowoomba Grammar School. During this time, wins various literary prizes.
1947–48	Works in Brisbane at the *Courier-Mail.*
1947–51	Writes poems and plays.
1951	Emigrates to England, where he continues to write.
1954	Returns to Australia early in the year, and later returns to England.
1955	Begins meetings with 'The Group', in London.
1960–68	Works in advertising.
1961	Marries Shirley Jannice Henry. *Once Bitten, Twice Bitten* published.
1962	Katherine, his first daughter, is born.
1962	Poems published in *Penguin Modern Poets.*
1963	Poems published in *A Group Anthology.*
1964	*Poems Ancient and Modern* published.
1965	Another daughter, Jane, born.
1968	Becomes a full-time freelance writer.
1969	*A Porter Folio* published.
1970	*The Last of England* published.
1972	*Preaching to the Converted* and *After Martial* published.
1973	*Jonah*, with Arthur Boyd, published.
1974	Visits Australia. Death of his wife, Jannice Porter.
1975	*Living in a Calm Country* published. *The Lady and the Unicorn*, with Arthur Boyd, published.
1978	*The Cost of Seriousness* published.

1981	*English Subtitles* published.
1982	Death of his father.
1983	*Collected Poems* published. Wins the Duff Cooper Memorial Prize. Takes up a position as writer-in-residence, University of Melbourne.
1984	*Fast Forward* published. *Narcissus*, with Arthur Boyd, published.
1985	Takes an Honorary Doctorate of Letters, University of Melbourne.
1987	*The Automatic Oracle* published. Honorary Doctorate of Letters, Loughborough University. Takes a position as writer-in-residence at the University of Western Australia.
1988	*Mars*, with Arthur Boyd, published. Wins the Whitbread Prize for Poetry for *The Automatic Oracle*.
1989	*Possible Worlds* published.
1989	*A Porter Selected* published.
1990	Wins the Gold Medal of the Australian Literature Society.
1991	Marries Christine Berg.
1992	*The Chair of Babel* published.

INTRODUCTION

To judge by casual report, anyone opening Peter Porter's poetry for the first time is likely to find it a challenging experience. It is often hard to say precisely why a particular poetry is in fact challenging. It may be a matter of esoteric allusion, of originality in approach to the subject, of intense address, of sheer copiousness. It may be the poet's discernible conviction that poetry is not at the service of other ways of speaking, nor even an alternative to them, but is a primary articulation of whatever it is that we are all trying to mouth during our lifetime.

It is a common view that poetry would be something else if it could—perhaps something more elevated, or more practical, but at least something less strange, less distinctive. Kinds of poetry which unabashedly and impenitently go on being themselves constitute challenges by that very fact. Randall Jarrell wrote once, of death, 'I feel like the first men who read Wordsworth./ It's so simple I can't understand it.'[1] Good poetry, first read, is so *something*—if not simple, at least steeped in its own selfhood: a new way of being.

This is certainly true of Porter's poetry. To introduce this small book on his work, I want to begin by looking at a couple of his poems, not attempting to catch their air or tone to any great extent, but simply pointing to some of their features, features characteristic of much of Porter's work. Both poems are taken from his most recent book, *The Chair of Babel*. The first is called, 'The Child at Sixty'.

> The harpsichord plays out of black or silver grooves
> And undulant air receives the message of
> A past where with the rhythm of hard hooves
> Each inexorable part of you vibrates
> At memories of gardens, faces, loves,
> The hiding behind doors, the padlocked gates
> And backs of friends, the well-invested hates.

Such fear and palsy in the grey will then compel
The greedy child to claw back sixty years
And find himself along the path from hell
He missed first time, and being old but now
Replete with charm and cunning trim his fears
With bright armorial lies and mutter how
What time condemns the music will allow.

TCOB 2

The first thing to notice, as ever, is the title. There are four things about it that I would mention. It is of course a paradox: there are no sixty-year-old children, though there are plenty of immature sixty-year-olds, and plenty of people who have descended into 'second childhood'. It is a common thing in Porter to give a special 'tilt' to experience, to angle-in on what he is watching. If it is now a commonplace that we see everything 'as' something, see it in a certain light or under a certain horizon, it is true to say that Porter often makes that 'seeing-as' explicit. He can, as here, do that by overthrowing expectations and subverting the matter-of-fact. Not everybody, reading one of his poems, will be prepared to be convinced: but it is a good working policy that one should be prepared to be surprised. Time and again, in the poems, one will be reminded of the Metaphysical poets and of those who have been their beneficiaries when it comes to seeing the world with a startled freshness. With some poets, their manner suggests that they wish to be the unnoticed mediators of a notable but unsurprising world. Reading Porter is never like that. Metaphor is ubiquitous in his work: so is the traversing of boundaries between conventional poetic decorums: so is his own kind of dream-work. A poem of Porter's does not just happen in the consciousness: it intervenes.

A second thing to notice about the title is its suggestion that the present endures the past, the past persists in the present. Hardly anything is a more emphatic influence in Porter's poetry than this sense. Sometimes the past rises up as the accuser of his present, saying that he is still irresponsible, or callous, or otherwise distasteful; sometimes the past brings good news, as the matrix in which treasured selves or treasured works present themselves. Good news or bad, though, the fact that we are time-kept beings, curious amalgams of the fixed and the fluid, is a major concern of Porter's. The re-doings of the portrait of the artist which can be seen throughout his work are not simply revertings to a convenient subject; they are the natural loci for the handling of a possessing theme. Later in this book I speak about the importance of metamorphosis in Porter's poetry; that is only one specification of Porter's immersion in 'the burning element', time.

A third feature of the present title is that it might, as well, have been the title of a painting. As such, it would have implied a 'study': something posed, addressed, and worked through intellectually. Whatever the lineaments of the figure thus depicted, an element of conclusion, of verdict, would be suggested. 'An element' is not all, of course, and some of the greatest of portraits imply at least a tension between conclusiveness and evocation in their rendering. In much of Porter's poetry one has some of that complexity. Part of his mind is in sympathy with design, structure, and finish. Often when he is writing about the music or the paintings which he admires—and reveres—the reader is conscious of high enthusiasm for accomplishment, for the magisterial. In conversation, and in his many broadcast or written commentaries on music in particular, the same trait is clear. With another part of his mind, though, he is a favourer of the mettlesome, of contravention and contrariety. Chesterton said about Dickens that all you could do with him was walk around him, cap in hand; Porter is more like Smart's cat Jeoffry, 'brisking about the life'.[2] Something in Porter wants always to be 'brisking' in the face of any subject, novel or familiar. This may show itself in tonal shifts within a poem, in curiosities of perspective, or in lunges of insight. A Latinist himself, he would know that '*studium*' not only means 'application', it also means 'zeal'. His 'studies' are commonly zestful.

The fourth thing I notice about the title is its implication of measurement. This is a little different from 'structure', in that it implies, to use one of Porter's words, the 'inexorable'. The inexorable is, literally, that which cannot be prayed-off, the destined or fated. This exercises Porter a good deal. We find in his poetry a resumption of the enquiry, as ancient as the oldest of Greek philosophy, into the relationship between liberty and determination in individual lives. 'Individual lives' it usually is in Porter, because although there is plenty of comment in passing, some of it lancing, about our shared condition, his imagination is centred naturally on individual fortunes and predicaments. It is in these, rather than in communitarian or political situations, that he finds whatever he takes to be truly representative. If he is a tragic writer, and he often is, this has little to do with society's inequities, and much to do with those parts of the human condition which are organic, individual, and endogenous. A natural and instructive comparison would be between 'The Child at Sixty' and Yeats's 'Among School Children', where the 'sixty-year-old smiling public man' poses to himself some of the same questions about inexorability that rise not only from the present poem but from many of Porter's. Nothing in Porter's poem is injured by the comparison.

The next thing I would notice about the poem is its structure. Porter, like many other poets in this century, has a hospitable attitude towards questions of structure. Liking formality, he dislikes rigidity; he has a very energetic intelligence, which is likely to be well served by discernible terms of reference, but which chafes under too-great predictability. In the present case, the poem is made up of two sentences, each composing a whole stanza, and with what might be called the ghost of a sonnet hovering in the background. We have, in short, conformities and deviations—or, as the poem might itself have said, variations on a theme. The book from which this comes, *The Chair of Babel*, includes poems with a good deal of structural variety, and a glance through any of Porter's books will show the same thing.

It is worth making a general point about those structures, even at the risk of sounding simplistic. This is that, at least nowadays, decisions about structure are always decisions about understanding. A polemicist like William Carlos Williams took this as a given: rightly or wrongly, he thought that various traditional measures encased and stifled imagination, and he hoped that by going a very different metrical way he would also be going a different imaginative way. Part of Porter's variety of structure is due to his constant hunt for a refreshed presentation of reality. At the same time, given his pronounced sense that reality has a strong vein of the anarchic to it, he likes those structures which mime constancy; they solace a condition which they can never adequately represent. Pascal says, '*Car la vie est un songe un peu moins inconstant*'—'Life is a slightly less inconstant dream'[3]—a saying which has the seemliness of an alexandrine in French, and of heroic pentameter in English, and in both cases has that seemliness keep less-than-obsessive faith with what is being said. It is the kind of gambit which would appeal to Porter—as would the sentiment itself.

There is an introductory note to Donald E. Westlake's *Good Behaviour* which goes, 'All of the characters, corporations, religions, nations, philosophies, political viewpoints, events and makeup in this novel are fictional and are not intended to portray anything in the world as we know it. None of the characters represents the author's personal opinions, which are, in any event, not quite yet organized into publishable form.'[4] Much in this would be sympathetic to Porter, even though he is not in the business of writing comic novels: certainly the drollness of 'not quite yet organized into publishable form', where 'personal opinions' are concerned, is something which he would welcome. And yet, the latent comic force, like the overt tragic force, of much of his writing becomes the stronger for the structures which cannot entirely contain them.

Next, music. I have little to say about Porter and music in this book, but that is largely for reasons of space. The subject warrants a good deal of attention from anyone seriously interested in his work. He has often pointed out that, in the arts, it is music that is his first love. The relevance of that here is that music not only provides him with a frequent subject, but is an implicit hovering metaphor, a sponsoring presence, in much of what he writes. Music has four roles: it is disturber, mentor, palliative, and transfigurer.

Music disturbs in Porter's poetry because, while it is 'of us', it rebukes us for all that we are not. It is coherent, while we are so only intermittently: it modulates, whereas we seem often only to mutate: it concludes, where we halt. Subject to its visitations, we are rarely entirely given to it: and whether we partly ignore it or turn it to our own purposes, we are often diminished by our response to its grandeur. When, at the end of 'The Child at Sixty', the 'child' is prepared to 'mutter how/ What time condemns the music will allow', that is scarcely music's word: it has its own inexorability.

Yet, it is a mentor. Porter reports that he always listens to music for a couple of hours before sleeping, and although this might be done in various spirits, it sounds like the action of a man who is, however congenially, being re-schooled constantly. In literature as in opera there abound figures who are manuductors, whether of the timorous or of the confident; these charm or sing their way into the confidence of the one who needs new directions or a new destination. For Porter, music is the great manuductor, and it maintains its own nature and character, unamenable to being rendered into other terms, and yet at least as 'relevant' to human policies or purposes as any other mode of communication or communion.

But the mentor also has his or her mercies, one of which is to palliate daily or lifelong distresses. As, in Christian theology, the Creator Spirit or Holy Breath can be stimulus or solace, so for Porter music is, for all its stringencies, 'on our side', a mode of grace. It is in a spirit of gratitude, as well as admiration, that he alludes so often to the makers of music and to the music made. It is hard, off-hand, to think of a single poem of Porter's which is not somehow about our woundedness, but music is the reliable physician of our ills—even though, no more than literal doctors, it cannot obviate our mortality. Another poet said to me once, when I was speaking lightly of trees, 'trees are the world to me'. Music is the world to Porter.

And finally, music has a transfiguring role in Porter's work. 'And Paradise, till we are there/ Is in these measured lengths of air', he writes.[5] No doubt he would agree with Proust that the only paradise we know is one that we have lost; and yet the place or condition goes on haunting him with increasing rather than decreasing force. He has often spoken of intuitions of 'otherness' or 'alternative worlds', and although he is firmly not a believer in any version of Christianity, the imaginative apprehensions which Christianity has often mediated are powerfully at work in him. Music plays some of the imaginative and emotional role for him that sacrament and miracle play within the Christian sensibility.

The second poem is called 'At Schubert's Grave':

> They took their calipers and measured
>   Dead Schubert's skull,
> So Science was by Music pleasured,
>   The void made null.
>
> What could that space of fleshly tatters
>   Say of its time,
> Of keyboard lords and kindred matters
>   Of the sublime?

The integers took up the story
        In fields of snow
And dreams through every category
        Were leased to go.

His was the head which notes had chosen
        To move within—
What gods and scientists had frozen
        Melted in him.

TCOB 63

In this poem we may see many of the elements I have been noticing, but others can expand our view. Firstly, it is a poem written in traditional vein in that it takes the spot made potent or memorable by the interring of a prized person, and lets the force of that person continue to discharge itself in the poet's imagination. Thousands of such poems have been written, and most if not all of them have been essays in the monumentalising of the person commemorated. This, it is worth noting, is a particular case of poetry's constant attempt to transform, transmute, translate: it is what might be called the 'real-ising' of the given.

Porter has written a number of poems which move him into the presence of the one who is none the less irrecoverably absent. The best known of these are those prompted by the death of his first wife, but there are others which celebrate the recovered presence of artists in their work. Porter's own view of the quest for 'the person behind the work' is that this is usually an act of vulgar confusion, in that the person truly to be prized is the person-become-work, whether this be Mozart, Wallace Stevens, or Caspar David Friedrich. At the same time, he is acutely aware of the constant traffic between our mortal days and the relative immortality which accrues to works of art. One feature of his 'tomb' poems is that they have something to do with that traffic; it is the *life* of the dead that is monumentalised.

By the same token, his acts of interrogation, or plea, or celebration, prompted by those Auden calls 'our good dead',[6] are also articulations of the common condition. It is as if death has removed distortion from the mirror in which we may see ourselves. The stilled dead still the living so that they may gaze without distraction upon the living and the dead. 'At Schubert's

Grave' is a representative poem of Porter's in that, attuning us to the exceptional, it also attunes us to the shared.

I stress this, because the poetry can on occasion be so imaginatively stringent, and so rhetorically oblique, or abrupt, that the reader could be misled into thinking that Porter's main interest is in arcane or recherché things. It is not. Those who value his poetry highly do so not because he can pull off amazing imaginative coups—as he can—but because he is trying to engage with some of the deepest, most disturbing, but also most consoling features of our common predicament. Amongst these can be numbered, on the one hand, our proneness to habituation and predictability, and on the other, our gift for novelty and creativity. Porter's poetry tries to give its due to each of these. There is a side of him which welcomes, and rests in, an eighteenth-century love of formal bearing: there is another which yearns for plenitude and upsurge, and relishes these wherever they are to be found. Any given poem of his is likely to carry the traces of both of these qualities. Saluting them in others, he enacts them himself.

'At Schubert's Grave' is a tauter poem than 'The Child at Sixty'. If the latter is disciplined rumination, the former is clipped meditation. The calculation referred to in words like 'calipers', 'keyboard', 'integers', and 'scientists' has its equivalent in the stanzaic arrangement, and in the wit that informs the whole poem. For instance, 'So Science was by Music pleasured' sounds like an expression from a seventeenth- or eighteenth-century lyric, but it gains force also from the twentieth-century reappropriation of 'to pleasure' from French literary theory; and 'What could that space of fleshly tatters/ Say of its time' works as well as it does partly because of traditional insights into our brief span's fugitivity, and partly because of collocations of space and time in modern cosmology.

But if it is a manifestly 'commanded' poem in these and other respects, and also, designedly, a more 'difficult' poem than 'The Child at Sixty', it grips partly because of its way of yielding itself to ungoverned and ungovernable things. I am thinking particularly of the third and fourth stanzas, where consciousness flows into the land of the uncanny. Dreams leased to go through 'every category', thus subverting philosophy's governing discipline: music making its choice of recipient, and having its will of him: these are imaginatively 'barbarous' and dionysian moves, pointing us towards realms unamenable to calculation. Science is an 'other world', and music is an 'other world', and we are fruitfully estranged when we are alert participants in them. But there is an otherness both within and beyond these, a mystery not reducible to problem nor calculable by science, even the science of musicology. That is where the poem takes us, without shedding respect for weights and measures. Again and again, in Porter's poetry, that is the effect one gets.

Another way of putting some of this is to say that Porter is above all else, imaginatively, a re-visionist, a revisor, of experience. When we read satirists or look at satirical cartoons, we find that the person writing or drawing has successfully pleaded the right to see things in a different way, to 'make strange'. The satirical streak in Porter is of this kind. When we look at a painting of Magritte's or read *Through the Looking-glass*, we see that liberties are being taken, but a liberty is also being created. Porter, in one poem, remarks that he has not called anyone 'sir' since his youth; his poetry does not call anyone 'sir', either—not even the great artists to whom he owes so much.

If there is anything to which, as a poet, he gives his allegiance, it is to the language itself. What the White Goddess is to Robert Graves, the English language is to Peter Porter. He is not unique in this, but he is unusual, even among poets. Jarrell hoped, though he did not expect, that critics would think good poetry more important than anything they could say about it; Porter writes as if he thinks that language is more important than anything that he can say in it. The undeclared surplus that is there in any of his poems is an act of homage to his medium. He is always writing, not at the grave of the language, but at its point of resurrection: his poems make up a zoology of the phoenix.

Porter gives the impression of going into language as the gifted dreamer goes into sleep. There are certain givens—memories, tropes, styles, fears, yearnings—but the realm holds enchantments which it will scarcely have begun to divulge in the course of even the longest lifetime. Ursula K. Le Guin has a book called, *The Word for World Is Forest*:[7] for Porter, the word for language is forest—something strong, complex, distinctively itself, 'other', partly domesticable, bountiful, and menaced. Porter's strong sense of menace in life is ubiquitous. He does not engage in jeremiads about the violations of language, but one can, very often, as in 'At Schubert's Grave', sense that wit is being used to retrieve language from inanition, and that an accomplished calm is being reached for lest language's meaning be chattered away.

To use terms little used nowadays in literary discourse, Porter's poetry is concerned for dignity, and even for nobility. Certainly, his eminently Australian scorn for self-aggrandisement has great play in his books, from the first to the latest. He can also have a Swiftian enthusiasm for getting his readers to think again, and feel again. And yet, the poetical quest is still somehow for that which is honourable, that which has an almost-Roman *dignitas*.

This, too, is a Shakespearean thing—the sense that, behind all the functionality, raffishness, squalor, lying, and mooning around of language, meaning, the unravished bride, waits. Poets in a score of modern languages have written in terms tantamount to these. As the globe packs, missiles are brandished, gabble intensifies, imaginations are sterilized, and hopes are turned to carbon, various custodians and beneficiaries of 'their own' languages have stood up to attest the vitality, in spite of everything, of those languages. Porter, I believe, should be seen in that company. He belongs with those for whom neither 'veneration' nor 'critique' is an adequate description of their attitude towards the milieu of words: 'unillusioned gratitude' comes into the account somewhere, and so does 'educated courage'.

'Books', wrote Paul Valéry, 'have the same enemies as men: fire, dampness, beasts, time—and their own contents.' True when he said it, it is still true. This present book is lessened by what it does not contain. It is, in fact, the merest primer to a reflective study of Peter Porter's work. It says nothing, for instance, about several of his most interesting books—*Jonah, The Lady and the Unicorn, Narcissus,* and *Mars*—each of which I read with fascination, a fascination redoubled by the collaboration with Arthur Boyd which they represent. Again, there are many unpublished or uncollected poems which are not referred to even in passing; and of course the vast majority of the collected poems are in the same case, so far as this book is concerned. Thirdly, anyone who believes, as I do, that Porter's prose writings usually warrant alert attention, whether for their own sake, or as a kind of Beatrice to his *Commedia*, the poems, can only regret that it has been impossible to do more than touch briefly on them.

Apart from the books of Porter's poems, and the many fugitive pieces not included in them, there are two indispensable resources for anyone seriously interested in his work. These are

John R. Kaiser's *Peter Porter: A Bibliography 1954–1986* and Bruce Bennett's *Spirit in Exile: Peter Porter and His Poetry*. Kaiser's book deals exclusively but very thoroughly with Porter's own work, and is particularly useful for its listing of uncollected poems in their places of appearance, and for the same service to the prose. Bennett's book, a full-dress study, gives us the life in which the poems arise and from which they diverge, and works systematically through each of the books to discern continuity and development. It achieves its aim admirably. Other writings on Porter's work which I have found especially useful are listed in the Bibliography.

What I have to add to things already written is a set of soundings of the poetry, and some reflections on ways in which it may be related to other verse. The approach is often impressionistic, but I hope that the reader will not find it gratuitous: it is always trying to keep faith with the poems. There is little that will pass for theory to be found here, but at this I do not blush: thought may do as well. My aim has been to leave the reader both more at home and more alert in the presence of Porter's poetry. A final chapter looks more closely at three poems, but its spirit is the same as that of the rest.

Andrei Sinyavsky, writing under his celebrated *nom-de-plume* Abram Tertz, has a line in *A Voice from the Chorus* which goes, 'Suppose one created one's own language and lived in it like an ape in a forest?'.[8] This magical notion is resumed in his novel, *Goodnight!*:

> I don't know why but, ever since the image dawned on me, I've always thought that a writer withdrawing to work and do some real writing retreats to the forest, so that no one will see him or hear him. With a single idea in mind: 'Time to put me on paper,' says the forest, becoming a text . . . the most important thing in the book you're writing is its mystery. For the author, for you. That is what excites you, what draws you in and quietly does its invisible work. That is all we need. And it's not important that people will argue and say, 'He made it all up, those things didn't happen'. They're in the fields, and we're in the forest. What they want is scope, scale, reproduction . . . But we are obliged to remember our refuge, our text, our secret paths. All they know is heroic epics and art's relationship to reality . . . And meanwhile all we have in hand is stories.[9]

It is not a bad epigraph for the work of Peter Porter.

1 The Province of Peter Porter

Books abound with titles like *The World of Shakespeare or The World of Dryden*: these tell us of the social milieu and physical circumstances and cultural currency which surround a writer. All of these things would have been there regardless of the writer. But there also exist books with titles like *Countries of the Mind*, whose stress is on something stranger. Here 'country' means both something which the writer inhabits, and something which inhabits him. If we spoke of 'Kafka country', we should be pointing to a landscape in which all is austere, menacing, exacting, and gratuitous. It is possible to establish connections of a kind, some of them causal, between all this and, say, Kafka's relationship with his father, or with the ambience of bureaucracy in the late Austro-Hungarian empire. But it would have been impossible to journey from those points in any direct way to Kafka country. His own imagination and sensibility, as distinctive as his fingerprints or voiceprint, preside creatively over the data from those other locales and, like God intervening upon chaos at the beginning of Genesis, make a world in their stead.

It is so with any writer of substance, and Peter Porter is eminently that. The metaphor of his own world or his own country is habitual with him. There are hundreds, perhaps thousands, of references to the pre-existing elements upon which he has intervened, and which continue to intervene upon him, in his poetry. Reading a typical Porter poem is among other things a process of reading off an abundance of items whose provenance is from his childhood or youth, earlier or later middle age. Their originating terrain may be Australia, England, America, Italy. They may be anything from a natural outcrop to an emotion recollected without tranquillity to a quattrocento painting. Usually, it is impossible to understand the 'pitch' of the poem unless one becomes to a degree inward with the character which these items acquire in it. In *Moby Dick*, Melville says of Queequeg that he 'was a native of Kokovoko, an island far away to the West and South. It is not down in any map; true places never are'.[1] Something similar can be said of Porter's 'places'—his Brisbane, his London, his Pisa, his Whitechurch Canonicorum. We can all go to places so named, but they may or may not go to us, and if they do they will not have gone to Porter-land. The same may be said of the wonderful array of impedimenta in his poetry. The labels on them may tell us that they are there by courtesy of Mozart, Rome, Ford Madox Brown, or the Australian bush; but while these are not falsifications, and while we often do indeed need them, they tell us little about what those items are doing in the poetry, or about what the poetry is doing to them.

In *The Place of Writing*,[2] Seamus Heaney dwells on ways in which locales, entering the poetry of Yeats and others, become metaphors of realities beyond themselves, as well as 'othered' confirmations of themselves. This is the kind of thing that happens very often in Porter's poetry. He often uses scriptural tags for his own purposes, usually ironic: he might appropriately quote 'my kingdom is not of this world', however much the London traffic may hum a few yards from the room in which he sits composing, and however often that worldly bustle finds an echo in the verse. I will have things to say later about 'otheration', and 'magicking', in Porter-land. For the moment I note that he himself, the 'I' of the poetry, is often registered as puzzled or disconcerted at finding himself in the country whose language the poems speak. 'What am I doing here?' is a question asked by millions of people throughout the centuries, about their dwelling for a little while on Earth; good writers, or some of them, ask themselves, 'what am I doing here?' about the country of the creative imagination in which they feel both at home and at best resident aliens. This, strikingly, is Porter's demeanour.

But he is no waif. There are plangencies and poignancies in his work, but in an obvious sense he is in command of the situation. Among the tropes which occur with some frequency in his poetry and his prose is that of saving things, or finding them saved, from the wreck. If he is a Robinson Crusoe in his own little land, he has Crusoe's inventiveness and resoluteness as well. It is worth remembering that a frequent method of composition for Porter is the highly traditional one of first blocking out the line of argument or progression, and then writing accordingly. This 'conduct of consciousness' or 'management of mind' is as important to the poetical outcome as the imaginative locale in which, again and again, he finds himself. I am reminded of the fact that in classical rhetoric the topos (place) is a familiar patch of mental ground to which the speaker has recourse, its familiarity being part of its value: but Porter revisits his topoi in order to rework them. 'Wrecked' though he may feel himself to be from some ampler and more befitting world, he is still mothered by present necessity into the reinvention, the rediscovery, of the land in which he finds himself.

Porter is highly alert to all this state of affairs. In many poems, he dwells on what it means to be here-and-not-here in the settings adduced for them. Titles like *Living in a Calm Country*—it is himself—or *Possible Worlds* signal what will turn out to be true: that he has an incorrigibly peregrine or exploratorial imagination. Places come over him, as it were, in the poems: he says, in 'The Last of England' (*TLOE*, *CP* 119), 'You cannot leave England, it turns/ A planet majestically in the mind'; and in the much later 'An Ingrate's England' (*PW* 3), 'You've given this country forty years/ Of your days, you're implicated/ In the injustices of pronouns/ And the smarter speech of sycamores.' Experiences reside in him: but he does not reside in experiences, or not wholly. In other poems, the steady epiphanies which he finds in paintings, or the shifting show offered by dreams, while being sources of fascinated meditation on his part, are also means of redrawing the borders between different zones of the universe which can be inhabited mentally. Many of his poems are about coming back—the poet's returning to a geographical or an intellectual or emotional topos: many other poems, and some of the same ones, are about things' coming back to him—in realisation, illumination, haunting or taunting. Though more sceptical than most about finding the answer, Porter is as interested as the next poet in the question 'who am I?' but 'where am I?' seems to be his defining enquiry at any particular moment. I am reminded here of remarks made to me by two philosopher friends on different occasions. One, when asked over the telephone how he was, said that he was all right, but didn't like the place much; asked what place he meant, he said after a pause, 'the world'. The other said that when you are lost you shouldn't ask where you are—you know where you are—

but where everything else is. There are elements of both of these answers in Porter's work.

In *All that Fall* Beckett, to whom Porter is indebted at least for the stone of unhappiness in 'What I Have Written I Have Written' (*ES*, *CP* 28), has a character burst out with the exclamation, 'Christ what a planet!'.[3] This is both pained and funny: the pain depends on worldly experience, the comedy on the intervention of a rhetoric which has given a tilt to that experience: the gambit, in fact, is Dickensian. 'Christ, what a planet!' might be one of Porter's atheist-humanist outbursts, and in a sense that is the way many of his poems work. It is as if the infernal or purgatorial investigations of Dante had been relocated in Surrey or Sydney: the writer, like the reader, has one eye on the brute intractability of the reality being characterised, and the other on the language which 'estranges' it. Lowell, perhaps borrowing from Baudelaire, says in one of his poems, 'anywhere but somewhere else':[4] a world elsewhere comes to be redefined by Porter from poem to poem, since the mind mutates constantly, giving a new othering each time round. (His array of memorial poems for his dead first wife is only the most striking example of this.) Being 'implicated/ In the injustice of pronouns' may be a phrase prompted by the ingrained social discrimination of conventional British speech, but it also suggests the living character of language—the giving of paroles, and the keeping or not keeping of them as the imaginative situation changes from moment to moment.

I call this section 'The Province of Peter Porter' for three reasons. The first has already been articulated: Porter country, like Greeneland, is the milieu of a mind. Whether the writer has been consigned there, like Ovid in exile by the Black Sea, or has made it his own and reworked it, like Crusoe, it is now his true topos—it gives his poetry its essential topicality. The second is that the notion of a province implies alternative terrains. One may not, again like Ovid or Crusoe, be able to visit those terrains, but word of them gets through, memories endure, aspirations provoke. Porter's is not a titanising imagination; he does not aspire, as some other writers do, to substitute his imperium for that of others. To read his work is to see originality and singularity of performance, but is also to be signalled back at many points to the provinces on which his abuts. His is, in that sense, both a classical and a hospitable imagination: as Dante accomodates Virgil, is indebted to him, and then bids him farewell, Porter has grateful but unsubservient relations with many writers and many worldly realities acknowledged in the poetry.

He has been challenged for saying that poetry is a modest art; but that claim fits well with his general sense that poems face outward into a world from which they may in turn be faced by others than contestants or confuters. Porter, who has made his living for the most part by writing about the music or the writings of others, is unusually alert to the paradox that we are awarded our distinctiveness largely in conceding distinctiveness to someone or something else. Pascal, writing what came to be known as his *Provincial Letters*, took it that the more decisively he characterised the position of his opponents, the more telling his own assault would be—as in the infantryman's dictum, 'find them, fix them, fight them'. Porter, vividly opposed to various human forces though he can be in speech or in prose, is not usually a polemicist in poetry, at least in the later work: but he has a Pascalian sense, with little of the Pascalian spleen, that the drawing of borders is part of poetical definition.

The third sense of 'province' is that which implies responsibility. Crusoe, enisled, gardens and cultivates his little territory. He does this in order to survive, but also because it is the thing to do. The notion of duty comes to us in one version or another with surprising frequency in Porter's poetry. It is clear that, for all the glint of ironies and the recurring self-deprecations, the sense that the work of poetry must be done is central to the work itself. Karl Shapiro said that whereas for Lowell the poetry was effortful and demanding, for Jarrell it came as if he couldn't help it:[5] for Porter it is perhaps something in between. What we see in his work is above all the poetry of the paterfamilias—the governor of a tiny imaginative community, all dependent in their variety upon his care, some begotten by him, each having his or her own direction and destiny, and all hovered-about by energies, benignities and animosities over which one has imperfect control, but towards which one must show the attributes of those whom Auden calls 'the tender who value and the tough who measure'.[6]

I foist nothing upon the historical Porter in saying these things. My point is that the agency which shows itself in the poetry, and which for short we call 'Porter', conducts the affairs of the Provincia Porteriana as if both justice and mercy had to be shown to all that falls within its ambit. Once again comparisons with Auden are instructive, especially if we are thinking of middle and later Auden, and if we have in view his sense of both the consolations and the obligations of civility, and his distressed contempt in the face of the botch so often made of civilization's immense possibilities. Another thing they have in common as poets is the sense that while the venture of art is not amenable to articulation on any terms other than its own, it can still justify itself abundantly even when assailed by those who find it no better than a diversion, an illustration of their own theses, an approximation to something better, or simply a mistake. That English should be 'kept up' is a fine idea, but there is no way of keeping it up in the abstract; Porter's poetry is a remarkably ambitious way of finding and showing how it may be kept up in one of its transformations.

2 THICKENING THE POETRY

That Porter country is there is beyond dispute—at least if the poet himself is to be believed. Apart from anything else, we find scattered through the writing 'The Porter Song Book' (*APF*, *CP* 110), 'At the Porta Humana' (*FF* 9), 'Reflection on my own Name' (*OBTB*, *CP* 14), and a nod towards Kafka beginning, 'Somebody must have been telling lies about Porter' (*TLOE*, *CP* 141). There is an element in all this of liberal deferentiality, a 'so I see it from where I stand' which is to be liked or deprecated according to taste. There is also an element of gaiety, a touch of cabaret. And there is an element of resolution, the poet being determined to lean from his established position towards what is to be encountered elsewhere: the matrix being established, 'into the dangerous world I leapt'. If we think of a particular poetical sensibility as being filamental—a thin, greatly sensitive medium upon which experience can impinge and through which, more or less changed, it can be transmitted—we are looking to see how things can be thickened up. The notion of the thickening of the plot is a familiar one, but it may also be a teasing one, at this stage of cultural history. The skeletal structure of event may be dramatic, and yet needs to be fleshed out if it is to be persuasive. In such fashion, the great novelists of the eighteenth and nineteenth centuries worked their way towards plausibility. By contrast with much of this, their challengers have often dismantled the assemblages, as if the Victorian treasure-house were simply a place to go to in order to die in comfort. What has happened in the novel has happened too in poetry: a maximalism has been confronted by a minimalism, the whole process being there summarily in *The Waste Land*, and being rehearsed by gifted writers ever since. The dialectic is easy to identify if one thinks of the rhetorical bacchanal associated with Whitman, or the rhetorical Lent associated with Dickinson: the one thickens, the other thins, and all is enriched as a result. But this example rebukes easy conclusions, for Whitman at his best has a steely intelligence dictating to his tongue, an intelligence which dismisses much and licenses little, while Dickinson has a sweep of wit which trawls oceans into the dock of her formulations. When we think about Porter, we think appropriately about a writer who has both Whitmanian and Dickinsonian traits.

He is like Whitman in his sense that everything is, potentially, apropos. The contemptuous account of Whitman is that he could not stop taking things in; the genial account is that all he registered pressed upon him as the child in the womb presses upon the mother, enjoining delivery. Whitman's poetry is various in quality, but he usually writes like someone in whom the flame of transmutation is unkillable. The sense of abundance in his poetry is due not just to itemising, but to the combination of lyricism, lexical originality, and the ubiquitous presence of will which encompasses the items. The Whitman who loved opera is the same agent as the Whitman who loved the scenic, the temporary, and the permanent. In his poems it is always the plot that is trying to thicken from noun to noun, from verb to verb, as well as the items or events that are trying to find a lodging. A particle-theory of Whitman will not do unless a wave-theory is added: the energies and accomplishments of his poems require both. Short poem or long, it is momentum as well as the monumental that is on offer in his writing.

Porter is like Dickinson in the sense that succinctness is a way of instituting reverberation. She took some of her metrical designing from the hymns then current, and perhaps she found in them, however much she transformed it, the intuition that whatever can be said aptly about life's great matters cannot be said terminally. From this it does not follow that things should be added, only that what is said should facilitate reverberation rather than muffle it. She too, in another mode than Whitman's, 'thickens' the situation, makes it more complex. In Porter's poetry, anyone will quickly identify some of the Whitmanian appetite for the nominable, and some of the Dickinsonian taste for the summary. But one would miss a good deal if one did not observe as well those complex resources and effects in his poetry which correspond to what I have been saying about these two poets. On a very small scale, his 'The Emperor Hadrian' (*TAO* 63), with the celebrated 'Animula vagula blandula', offers these things:

Little soul, like a cloud, like a feather,
my body's small ghost and companion,
where now must you go, to what region?
Pale little, cold little, naked little soul
who will you play with, what will you laugh at now?

Though he is controlled by his original in much, Porter (as his *After Martial* shows), is not held to the precise demeanour to be adopted, nor as to compaction or expansion in his rendering. What he gives us is a version in which denomination, suspense, and reverberation play together perfectly. Whitman is not the only one who has been going to operas, and Dickinson not the only one who has been remembering hymns.

I find useful here a line from Yvor Winters's 'At the San Francisco Airport', in which he speaks of 'the rain of matter upon sense'.[1] For Winters that is a tragic experience, and certainly for Porter it can often play a tragic part. At present, though, I use the expression more neutrally, to refer to that torrent of intelligible items of which, sensate and intellectual beings that we are, we may or may not make sense. In Porter's poetry we see an interplay between the rain of matter—that long tropical thundering down of the day's and a civilization's occurrences, of the flickering of sensibility, of stability and manoeuvring in relationships—and sense—the interpreting heart, the beguiled imagination, the sometimes ruthless judgement. Reflecting on this business, I classify elements of it, with a certain arbitrariness, into external itemizing, personal air, and synthesising vigour.

The first of these is inescapable. Run an eye down the first page of the 'Contents' of the *Collected Poems* (1983), and you will see *inter alia*, 'A Giant Refreshed', 'Syrup of Figs Will Cast Out Fear', 'Metamorphosis', 'Beast and the Beauty', 'Lament for a Proprietor', 'John Marston Advises Anger', 'Jack and Jill', 'Annotations of Auschwitz', 'Who Gets the Pope's Nose?', 'The Conservation of Energy'. Most people reading any of these titles will find an allusion immediately available, but in most cases there is a deflection of some kind. Porter, to repeat the earlier metaphors, trawls and farms: he takes and nurtures and readjusts the given. I cannot think of any substantial poet upon whom 'the rain of matter' drums in more unremittingly than things do for Porter.

Poets decide, by a certain stage, whether they are going to be invaded or to be invaders, or whether to be both. Porter has made the last choice, at least implicitly. Almost any page of his work shows this. Here, for instance, is the last stanza of 'Essay on Dreams' (*TAO* 38):

> Round the promontory, bursting like a cake,
> the ferry comes and soon its wave runs up
> the slimy ramp. Perhaps this is the spot
> where everything began; it could be we
> can leave the toys out on the lawn and start
> for home. Why not repeat yourself, the noisy
> insects in the trees assert—all good is
> habit that invades the homeless mind.
> And dreams have never heard of history
> or style, but like our childhood games
> they knock us into love with present fear.

Five stanzas have preceded this one, each of them itself an exercise or essay in dream-tokens and dream-styles. The topic is one which fascinates Porter: a draft of a talk given many years ago reports that he likes to think of himself as 'if not the Shakespeare, at least the Cecil B. De Mille of dreams',[2] and the mockery does not annul the truth in the claim. The first line of the poem is 'To set the impossible as homework!'—itself a reminder of the confluence between the near-majestic performances of the great essayists and the domestic ineptitudes which have become their sequel at the hands of labouring children. The whole poem makes the most, as dreams themselves commonly do, of the mingling of decorums which, in other circumstances, we assume to be styles of proficiency, but which are as arbitrarily indwelt as dreams. The movement through the poem is a movement from adage ('Training for death is what we do') through directive ('Look to the island where the herd of goats/ is from a shipwreck and every dangerous/ bird is widowed'), narrative ('Daylight drags itself through windows'), lament ('The great poems you could write/ are all assigned to Dickinson and Donne'), and tart jest ('Peculiar to hear "a person so obtuse/ he could not even dream us a straight line"'), to the beginning of the final stanza. There blossoms once more, as in most of the earlier parts of the poem, event after imaginative event. Dreams are, under some natural description, both things which break upon us, and things which we do—our passion and our action, our burden and our work. This piece of dreaming gives us pictorial vividness—the rounded promontory, the ferry as bursting cake, the wave gone up the ramp. (Perhaps it also licenses a conventional Freudian interpretation of such a moment, but it makes nothing of this.) Other sensations are there in the leaving, the noise and the knocking. There is the interest in aetiology ('the spot where everything began') and in destination ('start for home'). I think that there is an echo of Auden's plea in 'In Praise of Limestone'[3] not to be made like the beasts who repeat themselves. There is another of Porter's great cloud of references to dislodgement or homelessness. And there is one of his frequent exercises in dismissal of one characterisation in order to make room for another, which does not lack its own riddles; two stanzas back he has told us, 'and dreams are no one's coded messages,/ merely the second life our flesh is fed': now their task is to 'knock us into love with present fear'.

Whether or not Porter likes 'the place' much, it seems undeniable that he likes things. This is not to be taken for granted in a poet, any more than in anyone else. When William Carlos Williams's 'no ideas but in things'[4] achieved a favourable currency, the general assumption of the favourers was that one of those terms was self-guaranteeing, but it was not so then and it is not so now: plenty of readers, and an alarming number of writers, dislike things, not least themselves. By contrast, with Porter any number of things loom before him, soliciting approval or allegiance, but requiring discernment. The poems are both the memorabilia and the monuments of this process.

That being so, we are looking at the attingency of the world: its hooding us, embowering, englobing: its being not only world-for-us but arena of us-for-world. In either of these, surprisingly enough, we are looking at a continuing of the project of romanticism. The talmudic notion that 'the world is a wedding' has a variety of degrees of plausibility in everyday life, and has an insinuating force where poetry of the order of Porter's is concerned. The weddable are also the divorceable, and poets may come to say with Frost that they have a 'lover's quarrel with the world':[5] but even in that dire event a redefinition of the world is implicitly taking place. As well-calculated atheism redefines theism and deity, well-calculated quarrels with the world redefine the thing being rejected. Poetic disaffection always, at least obliquely, serves a cosmology.

Which brings me to a second point about the 'thickness' of the poetry—the matter of personal air. It is of course possible so to design poetry as to tend to abolish personal presence from it, or to seem to do so. This may be attempted because a critical theory requires it, as would seem to be necessary for any orthodox deconstructionist, or as a kind of *acte gratuit*, like writing a novel without the letter 'e' in it. But the idea has not taken with Porter: he is still, as one of his lines goes, 'Keeping ahead of death and Deconstruction' (*TAO* 6). The personal abounds in his writing, and may mean those things which we see to have their origins in diurnal event, as in, say, 'The Sadness of the Creatures' (*TLOE*, *CP* 136), or in 'An Exequy' (*TCOS*, *CP* 246). It may carry the trace of old occurrences, such as those of childhood, of which Porter has spoken or written at length.[6] But 'the personal' has another and subtler meaning. It has to do with what we call 'voice', the thing that in its literal contexts can signal both imminence and distance, the presenting of the person and the distancing of the person.

What holds together a poem of Porter's, along with those resources more readily amenable to appraisal like rhythm and (sometimes) rhyme, argument and association of images, is the sound of a voice which, with great mobility, modulates its way through the whole range of intellectual comportments, from certainty to scepticism, from the austere to the adorned: and which in doing so acknowledges a corresponding range of emotions, some of them enjoyed, some sought, and some avoided. That is how personality is shown, after all, outside poetry: personality doing nothing at all, and geared for nothing at all, would be unintelligible to us, and unidentifiable as such. There is no moment in Porter's poetry when, so to speak, we have the poet *pur sang*, but that does not mean that he is not there. And to a higher degree than is customary with many other poets, his dealing with the elements of his world is energetic. Someone said unkindly of Ezra Pound that his poetry looked like the china-shop after the bull had been through it; a poem of Porter's looks like the history of ideas after Isaiah Berlin has been through it—there is more energy around, more ideas around, and more order around, than was the case before. In his hands a poem intervenes, invests, jests, flourishes, enjoins, imagines, interrogates, and defines; although the moves are usually unpredictable, they are scarcely ever systematically or perversely so. They pull us much more firmly in the direction of romantic characterisations of the poetical pursuit than they leave room for a counter-romantic model. The words may not come as naturally as leaves come on a tree, but they come as appropriately. Another poet,[7] chidden for, as an opponent said, making 'what he liked' of a particular system of belief, replied that on the contrary he was making what he could of it. Porter's dominant air in any given poem is that he is making what he can of experience, ancient and modern, and that *he* is making it. Elizabeth Jennings refers in one of her poems[8] to poetry 'with its largesse and its own restraint': it seems a fitting way of speaking about Porter's writing.

None of this means that he cannot, with ease, emulate his admired Browning and conduct a highly skilful dramatic monologue, with no undertaking of unmasking: 'Soliloquy at Potsdam' (*1962/3*, *CP* 41) and 'To Start a Controversy' (*PAAM*, *CP* 71), are examples of just that. But listening to these poems we know within a few lines whose voice is really being heard, as we know Olivier's voice whether he is playing Lear or Jimmy Porter. It sounds, more or less, like this:

> Stella had a journal
> written for her
> the star of the world
>   our Stella
> had a journalist
> write about her
> when she got off the boat
> in mid-crossing,
> the water
>   a field of stars

'Real People', *TLOE*, *CP* 149

or like this:

> The oven door being opened is the start of
> the last movement of Rachmaninov's Second Symphony—
> the bathroom window pushed up
> is the orchestra in the recitative
> of the Countess's big aria in *Figaro*, Act Three.
> Catch the conspiracy, when mundane action

borrows heart from happenings. We are surrounded
by such leaking categories the only consequence
is melancholy

'Dejection: An Ode', *FF* 30

The air in both cases is that of a man who, however beset by burdensome or sombre events, is devising a way of appraising them. Irony is so constitutive an element in Porter's intelligence that he rarely has to refer to it expressly, but the mind is cocked at almost anything that comes his way. Few things, as he renders them, arrive with the Rilkean or Wordsworthian semblance of being 'the thing in itself': a light is being played upon them, and they are edged towards the centre of the page in order to be seen just so. This happens whether, as in 'Stella had a Journal', there is the semblance of another party's voice—that of a distressed parent—or whether, as in 'Dejection: An Ode', the music of what happens is the music of Porter himself in reverie. The first semblance is a put-up job, as surely as is Bishop Bloughram or Mr Sludge: who but someone with Porter's combination of the canted mind and the addiction both to Swift and to the ephemeral would have said or written it? Reading it, one hears fierceness and tenderness contending, just as they did in the *Journal to Stella* itself.[9] Swift mythologised Mrs Johnson in prose and in verse, played her as deity, wielded her imaginatively now as the sovereign of his own private star-system, now as his protestant equivalent of the 'Stella Maris', Mary; she was his subject, and he was hers. The drowned, perhaps insane, suicidal girl who bears the celebrated name has an entirely public memorialising, open to any amount of coarsening of the occasion, but when Porter is done with her she is exalted in more senses than one. In a way familiar in the modern period since Eliot, and ubiquitous in Donne's poetry and prose, the grand and the demeaned illuminate each other in Porter's poetry. Were he himself mythological he would be one of those beings—a centaur, a triton—who manoeuvre typically in more than one element. When one hears the Porterian voice, one hears tones invoking overtones.

One also hears a mind reflecting on its own affairs and fortunes. 'Dejection: An Ode' is one of many poems which announce themselves as artefact while preserving absolute liberty to melt the ice of art and take to the waters of outburst or outcry: others that come to mind are 'Essay on Clouds' (*TAO* 36), 'Farewell to Theophrastus' (*APF*, *CP* 116), 'Story Which Should Have Happened' (*PTTC*, *CP* 162), 'Sticking to the Text' (*TAO* 6). 'All I have is a voice', says Auden, at a famous moment;[10] Porter, like Auden, has in one way much more than that, a trove of experience both personal and vicarious; but what he has above all is a voice, modulated as the agent of the mind from the demotic pitch of oven doors and bathroom windows to the grandeurs of Rachmaninov and Mozart. The eight and a half lines which begin the poem do so under the aegis of the title. Coleridge, 'an Archangel, a little damaged',[11] looms over all, one of England's greatest intellectuals as well as poets, but always vulnerable to the self-inflicted humiliations which no one envisaged more accurately than he did. Walter Bagehot,[12] like many before him, saw the intrinsic oddity of flesh and blood being inspirited. Odd though it is, so it goes, and any poem of Porter's is the trace of its going, the trace of how it goes. Auden's choral figures in *For the Time Being* say, 'There's a Way. There's a Voice':[13] in Porter, the voice is the way.

It is a frequent characteristic of Porter's poetry, as of Montaigne's essays, that there is such a plenitude of memory, allusion, and reflection that one could think all this unnegotiable, indigestible. It does not turn out that way in Montaigne, for the reason suggested in the title of Jean Starobinski's *Montaigne in Motion.*[14] Anyone who remembers anything about the only begetter of this astonishing art-form remembers that he is the courtier of what he calls *une forme maîtresse*, which Florio renders as 'a swaying form'—something both undulant and authoritative, something intellectually erotic. Montaigne charms, but does so with an enchanter's powers: a Circean authority invests all of his discourse, so that one is aware less of items negotiated than of manoeuvres executed—of gradations and cassations. Something very similar is true of Porter's poetry. Those opening lines of 'Dejection: An Ode' have at least sixteen references to process in them: the form of the mind is a swaying form. In life, Porter habitually speaks with a bridled fluency: there is no verbal pottering about, no babble, and little, except socially appropriate, pause. Those qualities inform the poetry, in which, by contrast with much of his prose, he feels licensed to go all-out. The result can certainly be that the reader who, by custom or just this time round, comes to a poem like 'Dejection:

An Ode' with the expectation of imagistic stasis or measured unveiling or poised cadential effect will be disconcerted. Porter does not write free verse, but he does write libertarian poetry.

Three things open in the poem's first line: the oven door, a symphonic movement, and the poem itself. These are ontologically diverse, but rhetorically congruent, just as, elsewhere, a piece of music may encompass instruments, the human voice, and, say, a cannon. In fact, Porter's poetry is very often an act of orchestration: the effect offered is in part a schooling of one kind of alertness so that it may go with another kind of alertness. And those 'kinds of alertness' are also open to including kinds of emotional alertness. Emotions, at least in Porter, are not only fervours of the heart: they are fervours of the mind. So, here, 'mundane action/ borrows heart from happenings', and 'We are surrounded/ by such leaking categories the only consequence/ is melancholy.' The next sentence in the poem throws a beam of torchlight upon what has happened so far—'Hear the tramp of trochees/ as the poet, filming his own university,/ gets everything right since Plato'—and we are not surprised that it should, for already what has been generated is an alertness to anything that may come to hand, and what is literally to hand is the poem. But what does *not* ensue, and never does ensue in Porter, is a laying down of the arms of independence at the feet of solipsism. Talking the situation through, even if in soliloquy, is always a process of his talking himself more deeply into the world, rather than out of it. And the reader's sense of that world's 'thereness' comes, typically, not from the romantic *empressement* of rocks, stones and trees, but from its ability so variously to educe shifting alertness. In the desert you stare, in the jungle you scan: the world's jungle keeps Porter scanning.

I am speaking of 'personal demeanour', and of how it is elicited for the poetry and by the poetry. But it would not be accurate to suggest that we have in the end, in Porter, an interest in nothing but synergisms or inertial forces, of whatever order. As I have stressed earlier, he is *dinglich*, in his own special way. That 'oven door' may or may not, in the hands of the author of 'Annotations of Auschwitz' (*OBTB*, *CP* 30), have horrific overtones, that 'bathroom window pushed up' may or may not picture in part the filmed dash of many a trapped figure: but both seize domestic instants as graphic imperatives, things under the hand or the eye. Without such essays in the vivid, Porter's poetry would not only be the less adorned, it would be the less dynamic. Entities in his poetry are like the poles around which the slalom skier moves: take them away, and you take away most of the point of the exercise. But they are also like the landfalls on which the navigator, wittingly or otherwise, is bent: of the eighteen sentences which make up 'Dejection: An Ode', seventeen end with nouns. What pervades the whole performance is, after all, Coleridgean in more than its being a salute to that sponsor of another poem on melancholy. It emulates all that process of transmutation on whose behalf Coleridge may still be our best spokesman so far, the process whereby the 'shaping spirit' transforms event into symbolic registration—a tradition, if we are to go by retrieved artefacts, at least 300 000 years old. The reference to anciency is not irrelevant. Porter may not, typically, be as ready as Auden[15] or David Jones[16] to allude to the culturally primordial, but his instincts are aetiological, whether when looking at his individual fortunes, or when looking at our shared situation. The backward glance of the first five lines of his poem becomes part of the norm of attention. Otherwise, why should we have 'everything right since Plato', 'the Dark Lady', 'the beauty of dejection', 'to settle honey on my mind', '*Meermädchen*', 'I kept my father waiting'? Something of Porter's escapes us unless we see that he is, guiltily or gladly, in arrears.

I come now to my third point about the 'thickening' of the poetry—what I call its 'synthesising vigour'. Poems, like novels, essays, short stories, plays, are 'thickened' by the wedding and diverging powers within them. Porter's poetry does this with uncommon, and sometimes unremitting, force. All of us are the product of meiotic and mitotic powers, biologically speaking; things diverge and converge, whence our being, and whence, later, our style. In medieval western philosophy, it was said that estimation takes place *componendo et dividendo*, which has sometimes been taken to mean 'by the conflating of this element with that', but has also been understood as meaning, 'by saying yes or no'. Both stresses are appropriate when we are trying to understand Porter's poetry.

What has been said most recently refers to one version or another of conflation. There is a part of Porter's consciousness which is apt when anatomising is at issue—hence, for instance, 'Farewell to Theophrastus' (*APF*, *CP* 116), 'The King of the Cats is Dead' (*PTTC*, *CP* 155), 'They Come Back More' (*PW*

46). Taxonomy, tabling, come naturally to him. In that, he is at one with those poets who, as if originally or as if by profile against others, tally anything that comes to mind. Homer, Milton, and Whitman are only the most obvious. Seeded through the canonical writers are those who, given half a chance, codify as if their lives depend on it: Chaucer, Dunbar, Shakespeare, Pope, Wordsworth, Browning, Yeats. Poetry is a pendulum, and we cannot tell at what point the ring of meaning may resonate. Tallying is a way of making this happen. But Porter, like his striking predecessors, is not much interested in parade for its own sake. Coalescence and configuration are his real concern.

'A synthesising vigour?' When we speak of a theological style—Rahner, Schleiermacher—or a philosophical style—Wittgenstein, Locke—or an historical style—Taylor, Gibbon—we are always talking about a vigour of mind, which by turns embraces and sweeps aside an ensemble of items, and which affirms or denies certain appraisals as well-calculated to meet the case. We are talking in fact about connoisseurship as the governing principle of an intellectual discipline. And a Porter poem does exhibit in little an intellectual discipline. Forces external to the literary imagination have disvalued the possibility of its being of this kind, and have portrayed it as being at best a diversion or a muddle, a zone of perpetual immaturity: forces internal to it have done a good deal of damage by lending plausibility to the view that it offers at best a charting of emotions, things unamenable to structuring. But poetry which bears itself well is a refutation of these travesties, exhibiting as it does principled choices, apt conjunctions, and a gradually divulged design. It is in that sense connoisseurship; its elegance is the elegance of precise collocations. This is the condition towards which a Porter poem tends, whether the preferred mode be satirical, elegiac, warily celebratory, or something else. So, in the first stanzas of 'The Camera Loves Us' (*PW* 34):

> This is a lucky century, we have more
> To leave behind than just our bones.
> The minutes of our relaxation
> Cry for record, and where a spasm
> (Call it hope or pain) might once
> Have languished just in some sightseer's
> Slight recall, we have the lumps of light

To keep it fresh, to turn reflexes
To investments—finally to harrow Hell
With specimens of timeless platitude.

And that is just the start. The camera
Will paint us backwards. More than our
Disappointing mothers it will love us
And groom us into shape. It doesn't merely show
Us to the world but sprays us with its own
Invention till deep-seated ordinariness
Is lacquered by its gleam; the self,
Used to cracking knuckles in dark corners,
Is beamed out as a worldly figurine
Festy with the lens's afterbirth.

As often happens with Porter—'Too Worn to Wear' (*OBTB*, *CP* 34), 'Let Me Bore You with My Slides' (*TLOE*, *CP* 135), 'Exit, Pursued by a Bear' (*LIACC*, *CP* 233)—the title has a little course of its own, can make the rounds of its own topos. Certainly, on millions of occasions, people have courted the camera as if it did literally love us. This liveliness of theme is very characteristic of Porter. He enters areas which look apt for investigation, appraisal, and summary. The drive towards aphorism which can show itself in any circumstances has, often, a structural significance: the ground covered can be summarised. That is not to say that more has not happened than will go into any summary without significant loss: only that the intellectual rhythm of the poem allows for both discursiveness and provisional formulation. At the same time, the formula found may allow for, in fact call for, qualification or debate—as with the first couple of lines of the poem. The lucky century may turn out to have as ironic overtones as those called forth by the Lucky Country. That we 'can't take it with us' is to say the least not all gain; that we must leave behind our bones as if they were not ourselves is not a part of fortune. It is questionable already whether the camera's love will be unambiguous. The camera will be deployed as an instrument, through the poem, of investigation of ambiguity. The poet will point it, but it will be the thing carried for the purpose.

Which means that Porter is intent on finding something out. What is to be discovered, what uncovered, is the question that pleases and, sometimes, besets him. Memories, witticisms, observations, associations, emotions—all these crowd about him, and the discipline of the poem is a discipline of keeping as many as possible of these to hand so as to help constitute a world, while keeping a discriminating mind lively so as to attest to unsmothered selfhood. It is a significant and suggestive thing that in the second stanza he gives us 'world' and 'worldly' on the one hand, and 'our', 'us', and 'self' on the other. This is an apt comparison or contrast given what he is saying in the poem; it might also be the model for what is happening in many of the poems. A dialectic between self and world is not only the milieu of Porter's imagination, but a large part of his theme. This makes for frequent anxiety, since in spite of the very common references to 'we' in this and other poems, Porter's abiding sense seems to be that he is on his own in his investigations. His is a poetry of solitude, however many personages may be paraded through it, and however many treasured items may glow about him in the darkness.

At the same time, Porter's poetry is humanistic in an obvious sense of that over-politicised word; it works on the assumption that, whatever our fragilities and afflictions, we still bring abundant energies and modes of presence to bear wherever we may look or be. The petulant complaint, once common, that he loads his poems with cultural items pillaged irrelevantly, and somehow unpatriotically, from Europe misses the point that he sees all these things as ways in which the self has engaged the world, and so as resources for his own taking up the daily task, the daily address: the choice of them is a bid against existential loneliness and paralysis. Plenty of his poems go about that business in extended fashion—in fact there may be a higher proportion of them than even he would suppose—and even the present poem, in its remaining two stanzas, introduces Hadrian and Picasso. This poem depends on constant association with the high art of portraiture. And it assumes that what is of interest in either that art or the art of photography is not only, and perhaps not principally, the finished entity, but the drama of intelligence, expertise, and aspiration which is registered in the thing seen. Kenneth Burke's expression[17] for some forms of language—'symbolic action'—is a suitable one for Porter's sense of many a painting. In turn, his own poems offer symbolic action as prompted by the vigour and the coherence which he identifies as present or wanting in the scenes on which he turns his regard.

If in some sense his poems are 'speaking pictures', it is still the speaking that does the work. He has, for instance, a rhetorician's love for the sentence, something far from ubiquitous in poets. So, in the first stanza, after the couple of lines which could function as epigraph or as homiletic text, exposition opens out in one sentence for the rest of the ten lines. In the second stanza, the long sweep is delayed for a while, but once again it moves out to encompass items and perspectives. There is nothing automatic about this—it does not happen in either of the succeeding stanzas—but it is a suggestive move. It suggests a scanning command of what is to be seen or said, a mind spread as a hand may be spread to comprehend as well as apprehend. Elsewhere, especially in some of the earlier poems, the sentences are often short almost to the point of being telegrammatic, to signal tenseness or pain. In either case, we have utterance pitched so as to convey the sense that little is being done at random, even if much is said that will surprise. This too is connoisseurship, a discriminating alertness.

One of the offices of intelligence is to give pleasure, and a poem of Porter's is a structuring of pleasure. He has long been a student of earlier poets, whether as their occasional translator or as their beneficiary as a writer in the same language. He has also listened to and written penetratingly about his first love, music, and about its possible relationships with language.[18] It is to be expected then that his own poetry will have its own music or musics. At present I note only that these couple of stanzas give both elements of continuity and elements of variety to the ear. 'Lucky', 'languished', 'lumps of light', 'finally to harrow Hell', 'timeless platitude', have a euphony which is neither accidental nor beside the point; the caressing which becomes part of the theme of the whole poem has already begun without fuss. Miming, chiming, rhyming, the great staples of constancy in much poetry, are being given room to move here; it is an instance of what John Hollander calls 'melodious guile'.[19] At the same time, variety and disparity are given space: we do not merely look at, we listen to, say, 'It doesn't merely show/ Us to the world but sprays us with its own/ Invention till deep-seated ordinariness/ Is lacquered by its gleam'. This is play of vowels as surely as it is play of mind. Pope's preoccupation with dance of intellect to delight the ear finds a natural successor in Porter: wit sings its way in him.

At any turn, Porter's instinct is to associate and to discriminate. Open his *Collected Poems* almost at random, and this is what you may find. In 'St Cecilia's Day, 1710: In memory of W. F. Bach' (*APF*, *CP* 102):

> Having nothing new to say but born
> in the middle of a warm skill,
> watching the animals of art
> grazing undisturbed as if
> on the savannas of the moon,
> free by virtue of failure
> of the egregious disruptions
> of fashion, you are the star
> to hail across centuries
> of competitive rubbish . . .

The 'having' concedes association, and signals continuity; the 'in the middle' encompasses the given; the 'grazing . . . on the savannas' embraces stability. But 'new' and 'born' and 'moon' and 'free' and 'star' and 'hail' resist any constraining embrace; any definition given overall will have to accommodate vivacity. Porter's business is with setting and upsetting, both; he is donor and revisor, a renderer and a remediator. Neurotics do not have second thoughts: they have the same thought again and again, unstoppably. Porter has second thoughts in order to purify, modify, and sometimes to intensify the first thoughts. It does not matter to him whether he is dealing with a vulgarian—'*Unseasonable* knows a man who died three/ minutes after being diagnosed: it was only/ encephalitis but if he were my son etc.,' ('Farewell to Theophrastus', *APF*, *CP* 116), or with his own distressed self—'I offer my un-numinous body to the saint's care/ and pray on my feet to her merciful bones/ for ease of the ulcer of feeling, the starch of despair' ('At Whitechurch Canonicorum', *TLOE*, *CP* 138). Either way, we are dealing with a prevailing and countervailing intelligence, a resource for dissonance and a resource for consonance. 'Farewell to Theophrastus' begins, 'In a boardroom panelled with founders', and 'At Whitechurch Canonicorum' begins, 'This is a land of permutating green'. The terms of reference have been set from the first in both cases.

3 ROMANCE

Alfred North Whitehead describes the stages of learning as being those of romance, precision, and generalisation.[1] Romance is the stage at which we are all intrigued by anything presented for learning—are romanced by it, so to speak; precision is the coming to cases; and generalisation is the consequent informed reappropriation of romance, in outreach. As Whitehead puts it, '[The pupil] relapses into the discursive adventures of the romantic stage, with the advantage that his mind is now a disciplined regiment instead of a rabble'.[2] Whitehead's triad seems to me one of great interest and applicability, and to be useful when we consider the workings of Porter's poetry.

Undoubtedly one striking feature of a poem of his newly-read is that it romances the reader. How is this so? For me, one of the immediate attractions is the lexical sparkle. Joseph Brodsky relates of the poet Evgeny Rein,

> He once said the thing which I would normally say to any poet—that if you really want your poem to work, the usage of adjectives should be minimal; but you should stuff it as much as you can with nouns—even the verbs should suffer. If you cast over a poem a certain magical veil that removes adjectives and verbs, when you remove the veil the paper still should be dark with nouns.[3]

Reading Porter, one has the sense that whatever any magic veil specialised in taking off, what was left behind would glint with meaning. Sometimes this is due to comparative exoticism—'gerontocracy', '*Landsknecht*', 'femur'. More often it comes from the fact that, however much he may be engaged by words as his medium, and have a mannerist zest in setting them in the poem, he finds that each of them carries a charge of meaning which in part it brings to its context, but which in part it is lent by its context. So, to read, in 'The World's Wedding' (*TAO* 47), 'Speed is eloquence, rushing on to judgement;/ Nature the bardic never blots a line' is to be taken with an array of high-profiled nouns and verbs, but also to be caught by the sense that a matter of intricacy is being handled with summary authority. 'Rushing on to judgement' and 'never blots a line' are by now somewhat worn tropes—and aptly so, as the occurrences in dreams. They are revised, though, new-minted indeed, by their context; speedy eloquence is a novelty, and nature as Shakespeare a paradox. This refreshing of speech is one of Porter's great gifts, and one of the major attractions of his poetry.

In the same vein, he is as good a phrase-maker as they come. A rueful poem, 'A Philosopher of Captions' (*ES*, *CP* 279), alludes to this, and a man who has worked long enough in an advertising agency cannot but be aware that there is something questionable about the activity. Still, something of that mingling of instruction and entertainment which has its place in advertising comes up so inevitably in Porter's poetry as to seem second nature. 'Quartz studs glittering in a bank of clays' ('Going to Parties', *FF* 55), 'No tocsin of towns/ will resurrect you from the alphabet' ('Elegy and Fanfare', *FF* 27), 'The crinkle-crankle wall, the graveyard up for sale' ('An Ingrate's England', *PW* 33), 'The Fetishist of Polished Rhyme' ('The Story of U', *TAO* 62)—these make their own mark, assume their place, as apparently inevitably as melodiousness in Yeats or the whip-crack of wit in Donne.

We are 'romanced' in poetry by that which, when it happens, appears to have had to happen, as if the phrase or sentiment or trope had not so much been calculated to as been awarded to the writer—as if he had indeed been romanced by the Muse, and the poem was the consequence. For all the dexterity and the known designing, this is the look that Porter's poetry has to me. In a fashion very different from that of some of the Romantics, it is still a poetry of bewitchment, of beguiling.

It is unlikely that this would have come about unless certain topics or themes had a special sway over Porter's mind. Here, titles do not tell the whole story, but can be suggestive. Excluding the books done in collaboration with Arthur Boyd, I count twenty-seven poems with death or its equivalent in the title, thirty-one with music, twenty-eight with painting or sculpture, thirty-two with Shakespeare, drama or film, sixty-one with history or myth, and eighty-one with places. There are also forty-eight titles which include the continuous '-ing'. Sometimes these instances overlap, as in, say, 'Carlos Seixas, Dying'

(*TCOB* 64), or in 'Listening to Shakespeare' (*TCOB* 46), but that may heighten rather than lessen their interest.

Porter has often laid stress, in writing or in conversation, on the mind's elaborateness and ingenuity, and his own poetry is ample evidence for the truth of this claim, but like most poets he finds his attention polarised around certain matters, certain styles of being. The ones just mentioned are conspicuous amongst these. This is by no means the same as saying that he 'knows what to say about them'. His is, for all of its informedness and its cultural sophistication, the opposite of 'knowing' poetry at the level that matters. A metaphor used by Seamus Heaney[4] about the affair of poetry is that of divination, as practised by the water-diviner. Porter's poetry differs greatly from Heaney's in easily nominable ways, yet his too is an exercise in divination, if by that we mean a combination of precision in direction, alertness to stimulus, and confidence in the yield from attention. In a poem of Howard Nemerov's, the poet asks himself what he had wanted when he began 'that idiot fiddling with the sounds of things';[5] Porter's poems often carry the same insignia of self-interrogation, a poetical equivalent of the Wittgensteinian 'is this the way to put it? Is this what I want to say?'. They exhibit, that is, a striking combination of the gripped and the tentative. They have very little *raptus* or fugue, much engrossed attention, much controlled readiness for the next move. Here are a couple of cases in point.

> The evidence, like the weather, is from
> An inner storm; the poem, like every poem,
> Will be merely a beginning, the daring work
> For Venus's mirror, a heap of ashes,
> Never to be finished though complete:
> The mirror must be walked through
> And the one of many million crosses
> Borne—the pollen cloud dispersing, spruce
> Sadists and armorial villains stand
> Plausibly among scales and trumpets,
> Violence and punctilio masked by falling
> Almond blossom; then from the dream
> A voice proclaims the flag and day of evil,
> Indifferent to irony and liberal shame.
>
> 'In the Giving Vein', *PTTC*, *CP* 173

> But the power is still somewhere in us, hovering
> In the forehead auditorium of sounds:
> Those who were with us and have changed their shape
> Come back, like old ladies with parcels moving
> To the chair beside us; embarrassment abounds
> That pain is the one immortal gift of our stewardship.
>
> 'A Philosopher of Captions'

'In the Giving Vein' takes its title from one of the sadistic outbursts of Shakespeare's Richard III—'I am not in the giving vein', which God knows was true. In Porter's hands, this saying undergoes significant nuancing, so that 'give' can mean to make a gift, to cave in, to speak out. There is a sense too in which he is in the hands of the phrase; he is prone to listen to dicta as though they had a talismanic or a tributary force, a discharge to make upon consciousness, a begetting power; they can be anything from mnemonics to magic for him. What is going on here is an instance of what goes on in his poetry at large, the kind of thing indicated by the different senses of the expression 'a phrase', which can mean a couching of something other than itself, or (as in music) a couching of itself definitively. Porter listens for both of these when he listens to verbal fragments. The title of the poem before this one in the *Collected Poems* is 'On the Train between Wellington and Shrewsbury' (*PTTC*, *CP* 171), and of the one after it, 'A Hint from Ariosto' (*PTTC*, *CP* 174). Both make their bow to occasions beyond the poem's event, as does 'In the Giving Vein', but in this last case, the dictum invades the poem at many points, and once explicitly—'No pressure on my chest and yet I'm not/ In the giving vein'. The poem is, as they say, about writing poetry—it concludes, 'We're in a fire, singing; I'm the one/ Whose voice you can't hear; perhaps my round/ O is agony, I shall insist it's praise'—but is also about the constantly shifting relationship between saying and sayings, between the said and the 'said'.

Poetry is always about establishing, and also about yielding: about domination without domineering, about consenting without conceding. It is an improbable and a vulnerable art, because in one of its elements it exists at the behest of events, of which it is the registrar, while in another it constitutes the event, and is its own representative, is a patriot for itself. This is the milieu in which Porter moves consciously. Auden used to say[6] that the poet never knows, when he has finished one poem, whether he will be able to write another. Of course at one level this is untrue, above all if, like Auden and like Porter, one is a highly expert technician of verse; but in another it is invariably true, if what is at issue is that peculiar zone of the imagination in which poetry comes to be, a zone in which patience and initiative have so singular a relationship.

In 'In the Giving Vein', as so often elsewhere in Porter's work, dream is invoked, with its compound of the absolute and the relative, the firm and the fluctuant. No doubt it is here partly because he likes to write about dreams, but dream can also be a major metaphor for our waking ways, as well as a deliverer of messages, even if gnomically couched, for attention in another land. He is fond of introducing dream-passages into his poems, as if the dream-work were an evocation of the poetry-work always to be done. In the stanza quoted from 'In the Giving Vein', there is plenty of what might be called the bravura of dream, plenty of conductorly panache: 'The evidence, like the weather, is from/ An inner storm'; 'Violence and punctilio masked by falling/ Almond blossom': the maestros do it this way. But there is also trace after trace of insecurity and subjection: 'The mirror must be walked through/ And the one of many million crosses/ Borne'; 'A voice . . ./ Indifferent to irony and liberal shame.' Neither of these, imaginatively speaking, has priority over the other; this is not the denudation of an emperor, nor the muffling of nakedness, but the acknowledgement of both nakedness and clothes. If we ask by what Porter is 'romanced' when writing his poetry, and what capacity for romancing he helps those poems to embody, one answer is that it is poetry's capacity to translate our attention from one context in which completion and incompletion are mingled—the everyday—to another where they both have their claims once more, but on different terms. Reading a poem of Porter's is a process of tracing that translation.

'A Philosopher of Captions' works within a narrower compass—one of accommodation to the depreciating description of the work given in the title: it is 'being philosophical' about the limited predicament. But it is also, in its way, a more manifestly skilled or ingenious poem. Its four stanzas are organised around rhymes which have a controlled ungainliness, a good-tempered sheepishness: 'captains/captions', 'definition/fiction', 'text/wreck', 'their shape/stewardship'. They signal a sense that a miss is not as bad as a mile, that something is indeed being saved from the wreck, that flourish and polish are not all: they are philosophical after all. But this bringing of good news cannot annul the bad news, 'That pain is the one immortal gift of our stewardship'. Once again, command and vulnerability are interleaved with each other, are symbiotic. There are knowledgeable, even knowing, elements in the combination, but a chastening surprise is the predominating air. The ungainliness of those 'old ladies with parcels moving/ To the chair beside us' lodges in the consciousness as an element which cannot readily be dismissed.

Psychiatrists say that you can always make money by betting that the client will do it again. Porter's first book, *Once Bitten, Twice Bitten*, might have been titled as an illustration of this law. It is certainly true that typical behaviour, self-reiterating behaviour, is one of his staples of attention; he has a scientist's avidity for recurrence. But he also loves contrariety, second thoughts, and being a mental nuisance; he is in short an intellectual. His poems, like his prose and his conversation, abound in jokes; give him a patch of intellectual territory, and he will look for the fault-line. He writes in the stanza before the one quoted from 'A Philosopher of Captions', 'And the shouters, the ones met at stations by crowds,/ One can only admire them, join the acclamation/ And worry at their simplifying stance . . .' For Porter, it seems that there is a lot of shouting going on, most of the time, and his poems include a lot of worrying at that simplifying stance—worrying as distress, worrying as tearing back, taking issue. This too, I would suggest, is part of what romances him, part of what he transforms into romance for the reader: that the cultural given very often calls for confrontation and modification, not by the founding of some alternative school, and still less by *politique*, but by a more comprehensive and penetrating imagining. For all of Porter's inclination to articulate direness, there is something sanguine about this, a Wilburian instinct for the possibilities of bright junk, the dreck of things.[7] Dryden and Dickens have both been there before him in their sense that there is a muse even of garbage; every satirist knows that the worse things get, the more promising they are for the muse of the drastic. The interesting thing about Porter is that he has none of the afflatus of either Dryden or Dickens; he attends disasters with the air not of a magus, but of someone ruefully expecting imaginative gifts.

Which brings us to another of the things that 'romance' Porter—what Marianne Moore called 'the genuine'.[8] Having written a poem of some elaborateness on the theme of the worth of poetry, she cut it down to a few lines, which however still commend the genuine. Much criticism and much critical theory over the centuries has had to do with this matter, and it is I think one of Porter's working preoccupations. His is a hospitable mind, though he also has a hotelier's readiness to speed guests on their way; there are no fortifications around his consciousness, and little inclination to lock away the wilder residents. That said, he is a great one for scanning the credentials of what comes his way, and for offering alternative accounts of their identity.

Some of this propensity has become a commonplace; his 'On First Looking into Chapman's Hesiod' (*LIACC*, *CP* 210), and the debate with Les Murray and others which it sparked,[9] are well known. But the drive of which I am speaking is not confined to cultural arguments. Porter has a socratic sense that he is ignorant for the most part but always driven to become less ignorant. Being a man of democratic disposition, he attributes much the same combination to others, more especially those who have given themselves energetically to the making of art. If poetry begins in delight and ends in wonder, its path takes it through the vales of misgiving and critique, with the mists of falsification never far away. Porterian poetry, while highly susceptible to glamour and enchantment, still offers 'the examined life' as something of at least equal significance. Take, for instance, 'A Brahms Intermezzo' (*TCOS*, *CP* 268):

> The heart is a minor artist
> hiding behind a beard.
> In middle age
> the bloodstream becomes a hammock
> slowing down for silence—
> till then, this lullaby,
> arpeggiated thunder
> and the streams running
> through Arcadia. I, too,
> says the black-browed creature
> am in this vale of sweetness,
> my notes are added to eternity.

One great constant in Porter's poetry is the sense that, as here, death will have the last word, but quite as constant is his sense that he must, as an artist, find out what to do about this. In the *Collected Poems* this poem is preceded by 'Schumann Sings Schubert' (*TCOS*, *CP* 267) and by 'Non Piangere, Liù' (*TCOS*, *CP* 267): it is followed by 'A Scarlatti Sonata' (*TCOS*, *CP* 268) and 'L'Incoronazione di Poppea' (*TCOS*, *CP* 269). All of these touch, and some of them dwell, on death, and all of them invoke or insinuate dealings with it. Western cultures, like others, have historically been rich in recommendations and in techniques for addressing death, founded in biblical or in classical traditions, or elsewhere; everything from bracing oneself before it to embracing it has had a vogue, and continues to have its commenders. No doubt Porter like the rest of us takes from these sources whatever is to his own purpose, but his business as a poet is to generate a new resource, something which does not replicate what is already to hand, nor diminish the scale of the challenge.

What we get in 'A Brahms Intermezzo' is allusion to existing resources (the insight of *et in Arcadia ego*, the consolation of the music playing in the background) together with an essay in mingled decorums. It is an exercise in calling into the circle of the poem those other 'creatures' which may help to make a stand against the absurdity of death. Bloodstream, hammock-swing, lullaby, arpeggiated thunder, the streams running—all flow together as an intellectual and emotional resource against the undefeatable enemy. There is gallantry as well as ingenuity in the move. What keeps the poem entirely genuine is the irony of death's 'my notes are added to eternity'; the twelve lines which themselves constitute an intermezzo of considerable eloquence do more than concede, they proclaim, that the dark piper makes his own music. Whatever reservations one might have about Porter's poetry, one cannot deny that it gives the devil his due.

That is a grave kind of genuineness, and however frequent it is, it is not the only kind. Auden once distinguished between English and American kinds of 'fingering' in poetry, and illustrated this with a passage from E. A. Robinson's 'Ben Jonson Entertains a Man from Stratford'.[10] For 'fingering' one might read, 'distinctiveness of touch', itself a version of the style being the man. An alternative way of giving the devil his due is by shaming him, which is done by telling the truth; Porter's style is one made for the purpose. In 'English Subtitles' (*ES*, *CP* 273),

I suppose my feeling foreign is just because
I have never got used to ambition
In my friends—how can they think it worthwhile
Dazzling and adoring so persistently,
Waking up to find themselves half-famous?
Although I feel I am becoming extinct,
A sort of adaptation to unloveableness,
I also have my calling, the quest for
Freedom wide enough to die in. Thus spake the words,
Just as I was swearing to abjure them all:
'Perhaps you should say something
A bit more interesting than what you mean.'

Porter believes that the great differentiating thing between poetry and prose is the different ways of commanding the sentence, with poetry being called upon to conduct its affairs more subtly and variously. That is as it may be with other poets, but it makes good sense when one thinks of his work. His sentences are not only the vehicles of *sententiae*: they are the mediators of tone. Tone, the hardest thing to talk about without sounding like an over-ambitious describer of wines, is still of great importance where Porter's poetry is concerned. Just how one of his sentences is aligned against another, just how a phrase within one sentence is counterposed to those about it, is near the heart of his venture. Pace, pitch—these manner his matter, and although an inability to hear his poetical voice accurately would still leave the reader with plenty of quotable material, most of the reality of his poetry would be unavailable. Porter's art is, each poem round, a quest for the apt way of comporting oneself towards what is to be said. It is laden with all sorts of cultural allusion, which can be itemised at need or to taste, but while those items are often the objects of Porter's admiration, they are not his ultimate burden. His way of touching on them is an instance of his way of touching on one of his few great themes, and the *way* of touching on those is his project.

The passage quoted from 'English Subtitles' follows immediately Porter's saying that 'all the magazines I read,/ Those articles which make me reach for/ My spectacles spotting the word pluralism,/ Their substance is our beautiful Northern tongue,/ So useful for asking for fasteners in/ As well as for caning professors'. What he has to say, however pitched, will be part of the yield of the language, the gift of the tongue. Going back a little, like this, is always one's inclination when quoting Porter, because his writing, without being recessive, is constantly allusive to its own origins. Behind 'Those articles which make me reach for/ My spectacles' is Goebbels' 'culture' which made him 'reach for my revolver'; behind 'our beautiful Northern tongue' are celebrations from Herbert to Hopkins of the beautiful English tongue; behind 'Waking up to find themselves half-famous' is Byron's more agreeable fate; behind 'A sort of adaptation to unloveableness' is a mélange of Darwin and Freud; and so on. Porter would be a paradigm case for anyone wanting to press the claims of intertextuality, of language speaking us rather than our speaking the language, if it were not for the matter of tone, of voice. For the poetry does not sound, or feel, as if the lines were mere ensemble, the phrases mere echo. The sound of the 'I' is both modest and magisterial, as if he were, as he is, the gifted recipient of felicities.

'The genuine' in poetry may of course take countless forms, and each of Porter's poems is, to an unusually express degree, a quest for the apt form this time. The least of his problems is in settling on stanzaic form or its equivalent—the exoskeleton of the poem's embodiment of meaning. His is by now the proficiency of the virtuoso in such matters, though I am sure that he supposes himself to have much to learn there too. The problem is above all one of tone or pitch, imaginative dexterity, the timbre of intelligence. Socrates's celebrated daemon never told him what to do, only what not to do. Porter's sensibility is that of someone negotiating a way through immensely various imaginative possibilities, certain that there is far more than enough to be said at any moment, but picking his way through obstacles, seductions, mazes, dead ends. Earlier, speaking of Whitman, I mentioned the evidence of will in his writing: Porter's poetry exhibits much will, much tenacious purposiveness, in the midst of attendant and beleaguering possibilities. In that particular sense, it is a poetry of fortitude. This remains true even when, as with 'English Subtitles', the writing is veined through with comedy. That poem ends, to our surprise, like this:

As we sit in the dark,
I turn the more uncomfortable passions
Into slim sentences, such old words

As I know you know. This is language
I would go into the jungle with,
English of the Rider Haggard sort
Reversed-out on the glare of consequence.

Porter's being 'romanced' by the search for the genuine is something best characterised, as usual, through repeated reading of the poetry itself. Here, though, are several quotations from a book not designed with that poetry in view, but which seems to me highly pertinent to what he is about—Michael Frayn's *Constructions*.[11] (Throughout, paragraphs, not pages, are numbered.)

[48] Each of us develops a style—another attempt to order the chaos. Not by discovering an order in it, but by imposing one upon it, like a tank laying its own tracks across the wilderness. A man dominates his environment by establishing a unifying principle—himself.

[52] . . . Tentatively—as a joke, because it's spring—the oak sapling throws out small shoots all over the place without a second thought. But in time the tree has to follow in all seriousness these paths which it has laid down for its own development, growing massively into their capriciousness, and so taking on the fantastic gnarled shape of its maturity.

[178] Hold up your hand in front of you. Now give me a complete description of it. And you can't. Nothing that you can ever say will wholly describe the one thing that now lies in front of your eyes. You could describe away until the crack of doom. You could describe it as a whole, millimetre by millimetre, cell by cell, molecule by molecule. You could pile statement upon statement, and just as many more statements would remain to be made. And I should still not have available to me as much information as your first casual glance made available to you.

[233] Is there an order in the universe?
—There are orders. And disorders.

Frayn, obviously nobody's fool, is also nobody's servant; here, as in his novels, plays, and columns, he is intent on making his own way rather than on charting another's. But this handful of considerations seems to me to be very inward with Porter's policy and his *poiesis*, both. For he is, to use Frayn's expression, a 'dominator of the environment', of the environment as the milieu or matrix from which poetry emerges; he nowhere pretends to a Wordsworthian wise passivity, at least as the norm of his procedures. Equally obviously, the 'unifying principle—himself' is the thing being signalled at almost every point in his writing; a Lutheran '*ich kann nicht anders*' is the strophe of the oldest as well as the newest of the poems.

Still, in the same poems, and often in the same movement, his oak sapling 'throws out small shoots all over the place', and then pursues what might be called their bio-logic. A Porter poem is likely to be full of the controlled followings-up of possibilities which were originated in a way which could as appropriately be called pressingly as playfully. The thing to notice here is that the play of mind is not mere holiday from matters of moment, but is something to which any given poem's procedures appear to *call* Porter. Elsewhere, I have called him a *jongleur d'homme*, and that too is something which he 'cannot do otherwise', something which appears enjoined by the circumstances as decisively as the more astringent defining. 'Genuineness' of procedure calls some—a Rabelais, a Swift, a Borges—to denominate and to subvert, to create the rock-fastness of an island and its beach's friability. That, many hundreds of times, is what one gets in Porter.

If he were challenged as to why he should go about things in this fashion, he might respond in some such terms as those of Frayn's latter couple of paragraphs. Porter's universe is one in which plenitude vies with privation, array with disarray. For certain modest purposes—and satire's are comparatively modest—the superabundance of events and items may be tactically advantageous: that 'chaos is come again' is the habitual outraged warning of the classic satirist. And on the other hand, the designation of design suits very well the deepest instincts of the poet as such; apollonian or dionysian, he is still the singer of *tales*, the *magister ludorum*, the diviner of gestalt. But the truly ambitious poets are the ones who court resolution and rebuff with equal energy. Porter, lately, has been saying that now that he is no longer young he is taking risks in his poetry; my sense is that he has rarely done anything else, if we gauge the riskiness of the poetical venture by its openness to the affronting of one form of insight that another form of insight may bring. Wallace Stevens says, variously, 'Poetry must resist the intelligence almost successfully', 'Poetry is a pheasant disappearing in the brush', and (splendidly), 'The poet represents the mind in the act of defending us against itself.'[12] These are phrasings that might have been devised with Porter in mind. One reads him as one reads, say, Sterne: for the exuberance of the narration or the formulation, but at all the same points for the undisabled acknowledgement of being overwhelmed by the real.

If we continue to ask what it is that 'romances' poets, one answer must be in terms like those just given: that they are gripped by the opportunity to decide what they will endorse, while sensing that they themselves are claimed, are compelled. However they are gendered, they embody the gift, the giving, and the reception. This tensional condition is almost the definition of Porter's work. Take, for instance, a modest, middle-period poem, 'Occam's Razor' (*ES*, *CP* 280), and the opening of another poem from the same season:

> Never take the more unlikely explanation
> of any event in preference to the likely—
> just so, I say to myself, as I consider
> my life and conclude that this long haul
> to a sort of maturity is nothing more
> than a persistence of arriving in someone
> with nowhere to go, rather than, as I once thought,
> an inveterate but soon to be vindicated
> earnest restlessness at the port of life,
> a magic resistance to time, enough to
> keep me the youngest person in the room
> as well as the most ironically serious.
>
> But then reflect that if one is to tease
> out feelings from razors, one might prefer
> to think of Ibsen's madman sharpening himself
> or, in grateful suddenness, Haydn offering
> his English visitor that marvellous quartet
> in F Minor—what the gods mean by words
> goes the long way round or takes on flesh
> in dreams for the far vistas of nicknames.

* * *

> From the gunwale the dazzle is like spray,
> yet past the low reefs and their morning surge

gloriettes of cacti wait, summer houses
flensed from trees, a palimpsest
of childhood books and hardihood
of castaways. An hour ashore
and you forget your face, a week later
your colonists are raising ghosts
of the great world. Can all these
have swum in from the wreck—dreamers
stung by lexicons, lackland dowsers
imposing on the dark, sorters of bird droppings
granting interviews by gravesides?
Open the dictionary of discontinuity
and read about these islands.

'The Unfortunate Isles', *ES*, *CP* 308

Two sentences do the job in 'Occam's Razor', the first of them delineating a concession, an acknowledgement of the *dis aliter visum* in life, and the second seizing an initiative. Both of these are forms of endorsement, Porter's countersigning of what has proved to be so, whether in his own life or in that of others. They add up to an entry in the Book of Wisdom which he has been writing from the beginning, and none the less so for their citing, as does the original such book, many a peculiarity of behaviour. They are also, like all his poems, a kind of charm against time, something plucked from its teeth, and held up for admiration. And lastly, they make a joke to mark the world's jokes. William of Ockham might reasonably have expected that his paradigmatic rule for the conduct of science would have been dignified with the title of a law, rather than with a nickname; but the world is less amenable to bidding than his dictum suggests, and the best he could have was a joke. Porter's poem, from the first line to the last, is paying its dues to a necessitarian view of affairs, but is also preparing the way, taking it, or looking back on the journey into what can keep him and his reader agog. After Ibsen's madman, Haydn, and the anonymous visitor, who next, and what next, one feels—a sense heightened by Porter's own wide-eyed air in the second stanza by contrast with the demeanour recollected in the first. 'Far vistas' have been established, or at least provided for, by the time the poem ends, but its final word is a reminder of a governing which has been maintained throughout—and throughout a couple of sentences.

W. S. Merwin has a poem called 'Home for Thanksgiving'[13] which concludes, 'Oh misery, misery, misery,/ You fit me from head to foot like a good grade suit of longies/ Which I have worn for years and never want to take off./ I did the right thing after all'. I think of this when reading 'The Unfortunate Isles' and other of Porter's poems. Porter has in general far more esprit than Merwin, but they share a fondness for the absurd, a note struck at a number of points in 'The Unfortunate Isles'. Their terrain offers both the wondrous and the inevitable; everything about them is inverted, but stabilised in that position. This practice allows Porter an indefinite inventiveness, while insinuating that all has been foretold, fore-ordained, and gainsaid; it is one of Porter's many exercises in the infernal, but an infernal which has the sinister floridity of Hieronymus Bosch. '. . . Sorters of bird droppings/ granting interviews by gravesides/. . . aunts and uncles of the might-have-been/ distributing their malaise . . ./ Each is a creature calming himself/ with more anxiety': plenitude and privation, once more, contend for the honour of describing what is going on here. The fifteen lines which begin the poem are opulent with verbal flourish—'gloriettes of cacti wait, summer houses/ flensed from trees, a palimpsest/ of childhood books'—and with drollness, Porter himself being, obviously, a dreamer stung by lexicons, and a great reader in the dictionary of discontinuity. This is all very genial, in one way. But the violence of 'flensed', the vulnerability of 'you forget your face', and the perversity of 'imposing' all warn that, as A. D. Hope says of other islands, the rescue will not take place.[14] The simulacra of Hell are favoured imaginative territory for Porter, who usually treats Dante warily in his poems, but who goes the Dantean way of twinned revulsion and absorption when he writes of that terrain. If we ask once more what it is that romances Porter, the answer must be in part, 'Hell and the hellish in life'. But that nowhere forces him to deviate from his constant policy of taking pleasure and giving pleasure; he is the herald of a residual joy in the midst of disaster. The main form taken by the joy is the incessancy of discovery; even in hell, he is spoiling the Egyptians, as madnesses divulge meanings; 'perhaps my round/ O is agony, I shall insist it's praise' is not a line that belongs only in 'In the Giving Vein'.

The last sense in which I want to take the process of 'romancing' in Porter is his particular kind of fascination with the exotic and the beyond. Popularly, 'romantic' is often used in this sense—romantic Spain, romantic Greece—and 'the Romantics' themselves favoured such a usage. The post-Romantic Browning, if that is what he was, got much of this in 'a man's reach should exceed his grasp/ Or what's a heaven for?'[15] Goethe, whom Porter does not love, said that we never know what is enough until we know what is too much, and this is another version of the same thing. Communications engineers speak of 'redundancy', meaning not something which should be excised from a writing, but something which is logically unnecessary though perhaps rhetorically desirable. In that sense, much imaginative writing is a tissue of redundancies, whether we think of assonances, figures of speech, or the general organization of a piece. Such things go 'beyond' the palpably necessary, and whether or not we notice it they tend all the time to establish new frontiers of apprehension or comprehension: they are romantic. Cicero's writing is in this sense romantic, and so is Churchill's; but so is Dante's, and so is Hobbes's.

In Porter's case, we have a form of writing which, although he is not well-affected towards most of the Romantics, is 'romantic' in a plenary sense. Redundancy is everywhere in him; figures bud off vigorously; euphony or dissonance is organised into the affair with a generous hand; exempla for his theses (which themselves are multiple) arrive copiously. Even comparatively austere poems of his generate perspectives, nuances, innovations. Leavis, in a famously distasting essay on Swift, refers to his 'ant-like energy'.[16] If one can think well of ants, it is not a bad description of one element in Porter's writing; it is tirelessly spirited, incorrigibly energetic. True, what is being signalled is often the predicament imaged in an Indian prayer, in which the speaker refers to himself as going to and fro like 'an ant on a stick, both ends of which are burning'.[17] Porter very readily alludes to a hell on all sides, and there is no sort of poem of his which does not leave that possibility open; but while still on the stick, as it were, he is not to be stopped from envisaging superabundance in its many forms. I open at random *The Automatic Oracle* and find, as the first few opening lines, 'That same purr-voiced disc jockey' (*TAO* 2), 'There, in the re-grown jungle' (*TAO* 3), 'Adolescents smelling frangipanni' (*TAO* 5), 'In the Great Book of Beginning we read' (*TAO* 6), 'Where do we go to live? We're born ticking' (*TAO* 8), 'Flaking in a private room' (*TAO* 9), 'the last Muse is waiting on the corner' (*TAO* 10), 'The painted pale projectile moves and seems' (*TAO* 12). All refer to movement: all signal a 'beyond the present': all carry us forward, authoritatively, into what is to come in the body of the poems. Porter is, consistently and as it were instinctively, a Virgil to his reader's Dante, a guide not only to the ruins but to undiscovered cities, infernal lands, possible worlds.

But there is more to be said about his exotic, his beyond. If we think once more of Auden, it is easy to remember that, both in verse and in prose, he frequently delineated what might be called his nightmare world or city, and his dream world or city.[18] He could not go to either of these of course, but they kept coming to him. Porter's waking nightmare is that the dream world has been taken away from him; the Eden that never was ought to have been, and he ought to have had access to it. His mother's death and his subsequent exile in an odious boarding-school are referred to by himself and others as normative events in his emotional and imaginative life, and he has returned to these often in his work. But the sense of the 'ought to have been' can find expression at any point in the poetry, and the privation is not only emotional but metaphysical. Like the design of the Tower of Babel in Breughel's paintings of it, it is geared to go nowhere, even though it is highly geared. The 'horizon' under which Porter writes is the horizon of an unavailable paradise, and domesticate it as he may—'Paradise Park' (*TAO* 20), 'Paradise Artificiel' (*TAO* 19)—it will not let him alone. His many references to music are relevant here, since over and above their particular relevance in the individual poems (and the prose), they evoke our condition 'whose greatest comfort is music/ Which can be made anywhere, is invisible,/ And does not smell'.[19]

Chris Wallace-Crabbe has suggested to me that Porter's great good place is something like Freud's Rome, the place where the cultural fullness of the ages is simultaneously available. There is much in this. Porter, above all when he writes about Italy, sees villages or towns into which there crowd the consciously-appraised figures and items of the centuries. Sometimes, as in 'Pisa Oscura', this is done more or less systematically:

> Of the famed Piazza dei Miracoli
> This poem has nothing much to say—degree
> Of leaning from the vertical perhaps,

The bombs that made Benozzo's frescoes maps,
Incorrigible youth which carves its name
On G. Pisano's pulpit without shame,
Perhaps to discount in one heresy
The intolerable weight of history
And catch the tired tourist as he gapes
In wonder at the time-defying shapes
And with a CARLO or an ELVIRA
Restore the present in a vandal's scar.

TAO 70

Sometimes, the allusions are more arbitrary-seeming, and less predictable. In either case it is not the scanning of the scene that gives the effect of which I am speaking, but the sense that history will not just go on, and certainly not that it will go away, but that it delivers givens which it cannot thereafter take back. That 'time-defying shapes' speaks of more than Pisano's pulpit and its accompanying monuments; it speaks of the monumentalizing capacity of memory itself.

There is another kind of horizon under which all operates in Porter's poetry, and that is the horizon of death. Death, in upper or lower case, is named many times in the poems; it also wears many masks, and the unmasking is a frequent gesture. We are so accustomed to poems which speak, in many traditions, about death, that it is easy to forget that all of these are talking about a 'beyond'—not a life beyond death, but death itself. None of us has been there: in its respect, we are all like pre-adolescents thinking about sexuality. With Porter, one feels as if he has been there, not only in the vicarious sense that he writes often of the dead, but in the sense that he knows 'Zero at the bone'.[20] And yet it never feels like 'deathly' poetry, as, say, Kinsella's or Larkin's can. This is not because cheerfulness keeps breaking in, which it by no means always does, but because although death's horizon pangs him so much of the time, his sensibility is flooded with things which count against death. Death constantly reaches towards him, but it always has to reach. So, in 'Seaside Resort' (*PTTC*, *CP* 163),

With her nose turning from her shroud,
smelling of onions and Osborne,
the small black rained-on Queen

stares with disciplined eyes at the sea—
            between her and it
start and zebra-stop the export-worthy
cars, call the corpse-coloured gulls,
toe-tip and rock-a-nore the great
public and its implacable children.
            She can see fishing boats
and tankers, the only Imperial things
in her lawful eyes and even they
dribble tar like old incontinents
dripping it down their flies.
            She is more than a square—
she has all the Squares of England
and all the cats and hydrangeas at her call!
To be in two minds about her is right:
gone and for good riddance the Raj—
you ice-cream your mouth under her
undersized pedestal while the Eastbourne
Brass Ensemble showers its *Showboat*
Medley on the American-owned air,
but you regret Mr Mendelssohn
and the plant collectors of Udaipur.

'The polemic here is death', says the next line, death of an empire, death of company, death of the self. But while Porter may be doomed, he will not be hurried, and he will not be denied his own pacing the features of mortality. There is something in all this of Donne's mensuration of the *pompes funèbres* in his sermons and devotional writings; if we are the only species that knows of its own death, we do the knowing in our own way, and we call Death out on that. Porter, with death rarely out of his mind, makes as few concessions as possible to death, and calls it out as often as possible. His is, in fact, warriorly poetry.

Which is not to say that it lacks esprit even when dealing directly with death. 'The Melbourne General Cemetery' (*TAO* 53) begins,

This is a territory strange
To me, not the dead's embankment
But this southern city's range

Across a watered sky, the scent
Of burning, gravity's exchange.

My kept-neat city friends, the dead
I'll never meet, seem quite at one
With me. They are, no doubt, well-read,
Their stones and mounds dry in the sun
Following rain. They read in bed

Good news which goes on being news,
Old letters held before their eyes . . .

The spiritedness here consists partly in the jests, some of them on himself, but also in his tackling once more 'a territory strange/ To me', which is not only Melbourne, but the imaginative territory constructed by the poem, something as singular in its way as Stanley Spencer's Cookham, site for the oddest of resurrections. Good temper and quirk have their expression in this poem, but what I want to indicate is that it is typical not only in its negotiation with death, but in its 'othering' this place. Porter estranges places; they may or may not be geographically exotic—as it happens, I have a daily view of this cemetery from my kitchen window—but they are ontologically exotic. One of Ogden Nash's books is called *You Can't Get There From Here,*[21] and it is not only of the land of the dead that this might be said about Porter's poetry.

4 PRECISION

If precision is coming to cases, these abound in Porter's poetry. I pointed out earlier the large number of his poems which salute in their titles places real or imaginary. In fact, other places surface like volcanic islands in the poems; there is a not-so-hidden gazetteer in him. Take, for instance, a poem like 'Septimius Severus at the Vienna Gate' (*PAAM*, *CP* 47):

> Eastwards it is all plain with saw ice
> On the grass stalks; the cricket tinks.
> Fly butterfly, tell Rome
> The Man with the Nose is coming,
> He can smell pride as surely
> As a butcher's dog can sniff out a slaughtering.
> Tell Rome she stands on a southern sea,
> The winds from Africa bring her no good,
> They waft her an Emperor who speaks Punic.
> He will multiply religions
> Till they are as many
> As the ways men die; he will make
> Gods like starfish, stranded
> Prayer-wheels that point no way.
> Here by the Danube he swears
> He will fill the moon with death:
> The Emperor Semite
> Needs blood for the purple, let Rome take
> Peace from his hands: he has murdered war.

This is not in *A Porter Selected*, and is not mentioned in Bruce Bennett's *Spirit in Exile*; it is not what might be called one of Porter's canonical poems. But the moves recurrently made in it are typical ones, and one of these is his reference to places. Placing is like pitching, something at once hierarchical and exposing—disclosing, in short. The setting of the poem is partly European, partly African. Rome, already to a degree imperilled, is the nodal point of what it called 'the world'—much as, later, American soldiers in Vietnam would call America simply 'the world'. A larger arena still, 'the moon', presides over the scene. Within that scene, men and other living creatures manoeuvre, some naturalistically ('the cricket tinks', 'a butcher's dog can sniff out a slaughtering'), some bizarrely ('The Man with the Nose is coming,/ He can smell pride', 'he will make/ Gods like starfish'). The range of address is all the way from geographical observation, as in the first line, to a Tacitus-like ironical dictum, as in the last. Plenty is happening.

Within such a poem, the place-references anatomise and triangulate experience. In the poems of some others, we know place-names to be heavy with resonance, and to be talismanic entities—Yeats's Byzantium, Tennyson's Shalott, Marlowe's Ilium—and they may or may not have to operate in additional ways in the poems in which they occur. In Porter's poetry there is indebtedness to existing senses of particular places, but there is also an instinct for the revising of those places' meaning. His attitude towards places is one of dialectical conversation: he is their clinician, occasionally their pathologist, as well as their interlocutor. It is all the opposite of, say, Hollywood costume drama, in which we know the repertoire of streets, walls, trees, notices, vistas, and so on. Porter's attitude to places is like Socrates's attitude towards instances—they are more laden with intellectual and imaginative possibility when you arrive than they looked to be when you were making for them.

Three poems after this one, in the *Collected Poems*, is 'Vienna' (*PAAM*, *CP* 49), whose eighth stanza goes, 'At Whitsun cars and buses/ Carry large dragonflies, God can fly/ In this architecture . . .' Yes, it matters that this is the Vienna of *Gemütlichkeit*, of Mozart, of Harry Lime, of the *Anschluss*, as well as of the decisive battle for the routing of the Turks. What takes off from those matterings is the interplay or interflight between dragonflies and Pentecostal dove, the tussle between childishness and childlikeness. Vienna is being redefined in 'Vienna' as it has so often been in history. Similarly, in 'Septimius Severus at the Vienna Gate', the fact that we all know where to find Vienna on the map will not save that city (or the Danube, or Rome, or Carthage) from being relocated, re-pitched. Porter goes with precision to places in order, with precision, to be somewhere else.

We think of museums as places where the artefacts of the dead are displayed, but the name suggests rather places where the Muses muse. That is how places work for Porter. Some of his places are famed museums—Rome, Vienna, Florence, London—and each has its *genius loci* for him, to preside over its contents, but also to transmute them; and it is with that figure that he conducts his transactions. What happens when he goes somewhere, poetically speaking, is that he finds the locale to be nodal or nidal, but in any case a *point de départ* as well as a *point de appui*. He is an Heraclitean of place, as well as of time; it is not only nature's bonfire that burns on, but art's, the art of naming places. He has, as it were, invented a cultural Theory of Relativity for himself: time and space interlocute and mutually transmute. He has sometimes said that he has not, until lately, been good at looking at places, and perhaps that is true; but it must be hard to look at places when your imagination is moving faster than your eyes, and that is the sense that is conveyed by the poetry.

So how does he look at places? Like this, in 'The Cats of Campagnatico' (*FF* 41):

Not just a vignette, we reflect, this shadowless day
In Southern Tuscany, more a looking for shades
Which match the petrified intelligence of time:
One sees the small bends which history makes
In the lanes of scarcely-visited villages,
True, this one is in Dante, and that oleander-screened wall
You take for the headquarters of the Carabinieri
Might be an out-station of the Piccolomini,
If only you could remember which is which
Among the towers that mark the lesions of the sky.
Siena is as far away as London; life as far away
As last night's dream whose every promontory
Is in the present. Now, coming through the gate,
The view is a pastoral benediction for those
Who have never lived in Arcadia. *Thank God*,
Grace à Dieu, *Gott sei dank*—we are
As international as an opera festival,
We who love Italy. We have no home
And come from nowhere, a marvellous patrimony.

The consciousness is tugged out and pushed back, as the poet deals with place. It matters to him that we are in Southern Tuscany; Italy's fierce regionality is not something merely to be noted and indulged, but something to be seized as history's configuring of geography: we are pressed in upon locale. At the same time, and in the same place, things expand and pulsate; Dante looks after the universe as well as 'this village': Siena, London, Arcadia, inscribe themselves: the God who is traditionally the *patron* of all, whatever their language, is unavailable to offer a home, and the patrimony of vacuity takes his place.

The sculptor Michael Meszaros says, 'Every structure depends on the designer achieving a balance of the forces of tension and compression',[1] and it is a way of speaking which might well apply to Porter's handling of place. Place is rarely resting-place, and is usually the site of drama. He cannot be un-placed, but it is hard to be in places. In the first eleven lines of the passage quoted, there is not one which does not allude to deflection, alternative, or provisionality. There are plenty of allusions to the natural—'shadowless day', 'small bends . . . in the lanes', 'that oleander-screened wall'—but every time, these are part of a larger set of framings: the 'shades', 'history', Dante's poem, 'the petrified intelligence of time', 'the lesions of the sky', and the 'dream whose every promontory/ Is in the present' have as large a logical claim, and a more formidable rhetorical claim, on the attention. It was said of Tennyson that he was always carrying a bit of chaos around and turning it into cosmos; chaos or not, whatever accrues to a poem of Porter's tends to the condition of cosmos—and cosmos, while indulgent of places, is nowhere reducible to them.

It is obvious that Porter loves places, but he speaks his love in a different way from most others. Places attract from him an intricate subtlety of attention. Reading, 'Now, coming through the gate,/ The view is a pastoral benediction for those/ Who have never lived in Arcadia', one passes from the mimesis of transition, through a genial paradox, to an irony which first controls and then perturbs. All of this passage, it should be remarked, is a contribution to the effect of the poem, less 'charged', to my ear, than much of what comes before it, and some of what follows it. It is standard Porter; it is what happens to place when his Muse visits it.

We do not like things or persons to be deranged; we want them ranged and arranged. This goes for places too, whence maps—whose fascination far exceeds their utilitarian value. Porter has a geographer's readiness to cast around for a different perspective on the landscape, together with a cartographer's zeal to map what is then seen. The way he characterises places, factual or fictive, shares in both of those qualities. One grows so accustomed to one particular version of the disposition of locales that any other seems wrenched—not only, say, Steinberg's famous *New Yorker*'s view of the world as against Mercator's, but any at all not done on Mercator's projection, if we have become used to it. Porter's attitude to place is by contrast that of the young man who, rebuked for not following his father's ways, replied, 'He did not imitate, and I do not imitate'.

Archimedes said, 'Give me a place to stand on, and I will move the world'; Porter might say, 'Give me a place to stand on, and the world will move me.' He is not the inspector-general of locations, but the pilgrim of their possibilities. This is how 'The Cats of Campagnatico' ends:

> The cats of Campagnatico,
> Which are never fully grown and have never
> Been kittens, will not move for the honking motorist
> But expect to be gone round. Thin and cared-for,
> Fat and neglected, watchful and hardly seen awake,
> Cool-haired in the sun and warm in shadow,
> Embodying Nature's own perversity,
> They lie on this man-made floor, the dialectic
> Of survival. O God, we cry, help us through
> Your school of adaptation—between the fur of the cat
> And the cement of extinction, there are only
> Cypress moments lingering and the long tray of the sky.

Porter has done some poems expressly in the *tombeau* mode[2]—'At Whitechurch Canonicorum' (*TLOE*, *CP* 138), 'The Charnel House, St Leonard's, Hythe' (*TCOS*, *CP* 241), 'The Tomb of Scarlatti' (*PTTC*, *CP* 189)—but these are only a few of the poems in which place yields a meditation on mortality. 'Meditation' is an apt word only if it carries some of the traditional sense of an exercise leading to an arousal of the heart. It is important to stress, about Porter's poetry, that while he is of course disconcertingly well-informed about many things, and while he has both the penetration and the mobility of the real intellectual, what the poems seek to discover each time is 'the true voice of feeling'. He wants things to be intelligible, as far as this can be managed, but at least as much he wants emotion to be liberated and intensified.

If modern psychology has taught us anything, it is that the roots and character of emotion are much harder to identify than has often been supposed in the past, and that the apt expression of emotion is as much a matter of finesse as of spontaneity, at least for an adult. All this is true of Porter's poetry; that 'the truest poetry is the most feigning', or at least the most designed, seems to be an artistic axiom for him, even if given poems have a kaleidoscopic rapidity and virtuosity. What we have in fact is a re-presentation of the Metaphysical venture, as that was conceived by Donne: that mind and heart should issue organically in the poetry, and that nothing less than the whole person should cry out the whole person's needs. So, here, some cats, for Porter so often evocative presences, typify to him 'the dialectic of survival', which issues not in further appraisal but in a cry. He did not go to Campagnatico for this, but as happens very often in his poems, place does for him what persons do for some others—liberates feeling and exacts it. The poem began, 'Since a harebrained Devil has changed the world/ To scenes from a Nature Documentary', but we are a long way from the documentary by the time we come to the end.

T. S. Eliot speaks of places 'where prayer has been valid',[3] whatever exactly that means: certainly, there are places where, as we say, things *lodge* with us. This brings me to a second kind of 'precision' in Porter, the precision we identify when the sizeable lodges exactly. This does not come of finicality or preciosity, and it can take place in a number of ways. Richard Wilbur has a poem, 'Tywater', about a cowboy, which, after speaking of his skill with lariat and knife, concludes, 'The violent, neat and practiced skill/ Was all he loved and all he learned;/ When he was hit, his body turned/ To clumsy dirt before it fell.// And what to say of him, God knows./ Such violence. And such repose.'[4] Those last two lines could not have been educed from the ones that preceded them, being larger in scope than even the sum of all the earlier ones, but they lodge the meaning of all that has gone before them, as well as their own meaning: they are the point of decisive impingement of the whole poem. Lodging, impingement—these are salient in Porter's art. We can describe his attitudes and dispositions in a round-enough way, and he has often done it for us. But even poems of his which have a didactic element do not mainly display that element; one gets from them not illumination at large but sunrise or sunset, high noon or the crepuscular. He is a particular poet, which is not quite the same as a poet of particularity, though his poetry may give us both qualities, as both are to be found in 'Tywater'. Everywhere, in any case, he is aware of the sort of thing which he spells out in 'The Recording Angels: Music and Meaning in the Voices of Poetry':

> The hard rule is that poetry has to be loyal to language, to Stevens's 'essential gaudiness'. At this point I can hear the voice of George Herbert, a poet whose genius might be said to derive from his passion for truth—and then also of Samuel Beckett, whose progress was from an almost Yeatsian extravagance (I think Joyce is not to the point here) to the bareness of his last-breath last works. Yet what all these have in common is a falsification which is built into language, and each knows that beauty alone will justify their efforts—or, if not beauty, at least some sort of pleasure in the shaping of truth beyond its commitment to right behaviour. It is exactly if not precisely the voice inside the words which we believe in, and which gives us the satisfaction we call art.[5]

The 'loyalty to language' which is also a 'pleasure in the shaping of truth' presents problems, which are the subject of Porter's Warton Lecture, but taken together these attributes are to be found in what I call things' lodging with us. Examples abound in the poetry. Here is one to begin with—'A Study of a Bird' (*LIACC*, *CP* 218):

> With completely retractable sail
> he has an ocean surrounding him;
> his surface is at all edges
> and everywhere the reefs wait
> with their yellow surf: he lives
> in a cube drawn by a mad artist,
> his eyes are along diagonals;
> perspective is what we invent
> to be like him: he is not like us,
> he would never expect an epitaph.

Brave on a tractor while they lunch
he is as vain as a place god,
an inspector of three hundred tall
stalks and the ants' deep shelter.
That sound now—is it the frogs
in their safe swamp, or the crickets
down on their knees? No, it's
the bariolage of useless terror
made by all things born to die—
he dips his head, they are wiping their mouths.

Traction by Leonardo, costume by Audubon,
habitat by John Clare—the Establishment
of creatures chairs him to the top.
There he passes with his flaps down
plumb on a winter tree. We may
take his eggs and measurements
but we cannot levitate. A ring on his leg
will match him to God's census—
only his watching superiors leave
paradise, trailed by a flailing sword.

Fields at evening hold the honey
of the sun. He 'fleeth as it were
a shadow', the bird of the Angelus
among the gleaners of Nature:
it is all dark and light to him,
a scene by C. D. Friedrich to us.
Heavier-than-air machine the poet
loves, he overhears these pointless
poignant words: *I am a child of the*
Enlightenment, I expect to be happy.

The title sounds objective enough, and there are some cool-eyed elements in the poem, but consider how each stanza ends; the first three point directly to mortality in us, and the fourth to, at the least, reservation about the Enlightenment expectation. It is as usual the matter of the human that is being addressed here, as certainly as in any Romantic lyric or ode. But on the other hand, the bird is not just a property within the poem's drama; 'There he passes with his flaps down/ plumb on a winter tree' is a salute to him as well as to human ingenuity. The fact is that the poem's vigorous effect comes partly from Porter's engaged attention at more than one level within any stanza. 'Traction by Leonardo, costume by Audubon,/ habitat by John Clare' is wit rather than whimsy, giving as it does to each of those savants a fragment of precise description which has been transcribed from another context, and in each case to his advantage. 'We may/ take his eggs and measurements/ but we cannot levitate' acknowledges our competencies while denying us the fruits of one of them—the imagination of levitation. 'A ring on his leg/ will match him to God's census—/ only his watching superiors leave/ paradise, trailed by a flailing sword' puts him in the same case as the infant Christ subject to Augustus's census, makes mock of man's godlike pretensions, puts the 'superiors' out of paradise and into the market for epitaphs, and forestalls any easy assent to the poem's 'I expect to be happy'. It is among other things a bestiary poem, with the bird doctrinal *malgré lui* to its observer. Such is what might be called the ground of the poem, the milieu or matrix in which its various sayings lodge.

And how some of them do lodge! '. . . he lives/ in a cube drawn by a mad artist,/ his eyes are along diagonals;/ perspective is what we invent/ to be like him: he is not like us,/ he would never expect an epitaph'. It is like, in 'Sonata Form: The Australian Magpie' (*ES*, *CP* 283), 'You can upbraid the magpie,/ saying, "What do you know of Kant?"/ It might shift a claw an inch or two': definition potentiates definition. Porter's imagination works by cognates, isomorphs, disparities, gestalts, asymptotes. It is as if the medieval tropes of world as book and universe as music operate formidably when he writes. True, the book is dog-eared and water-stained, and has lost many of its pages, and a 'harebrained Devil' has got at the music, but what is left still licenses Frayn's 'There are orders. And disorders.' To acknowledge both the order and the disorder is very typical of Porter's project. His dicta—and many of his lines have that flavour—convey a sense that observation can both vindicate itself by its precision, and hold fast even in the shifting soil of the consciousness that receives it, first Porter's, and then the reader's.

All this, though, I would reiterate, is for the sake of feeling as well as of illumination. 'Mad', 'brave', 'vain', 'safe', 'useless', 'watching', 'flailing', 'dark', 'light', 'pointless', 'poignant', 'happy'—it is a little symphony of sentiment. The precision of the poem's wit not only makes room for emotion, but authenticates its plausibility: emotion lodges, rather than floats, in such a context. Elsewhere, for example in 'To Start a Controversy' (*PAAM*, *CP* 71) and in 'A Consumer's Report' (*TLOE*, *CP* 131), Porter makes play with severance between appraisal and emotion; but 'A Study of a Bird' is one of the many poems in which admiration for proficiency in an expert goes with a sense of the expert's, and our, exposedness to mortality. In most poems of his in which there is, in effect, a man in a white coat, the 'black-browed creature' of 'A Brahms Intermezzo' will be at his elbow.

'Study', says the *OED*, 'Drawing etc. made for practice in technique or as preliminary experiment for picture or part of it . . . literary portrayal of an aspect of behaviour, character, etc.': it is an operation in little. 'A Study of a Bird' is one of Porter's pastorals, something contained but far from tamed, a microcosm implying macrocosm. The scaling down can make for precision of touch, for engrossment, for heightened prizing of needle-pointed contact. He does this sort of thing often and well. Here are two examples, the first the whole of 'His and Hers' (*LIACC*, *CP* 198), and the second the end of 'Doll's House' (*FF* 23):

Mornings weaving through the mud
Gave his high boots a cracked glory,
And she, crying where she stood,
Heard the hardly bothered-with story;
Each not knowing that pain clings
To the cleaned surface of things.

The greenhouse overspored with seeds
And the motor mower dried of oil,
Over the garden virginia creeper bleeds,
The lawn has patches of bare soil—
This is the Estate Agent's concern,
Since he's in a box and she in an urn.

Behind a door missed by visitors
High boots and binoculars stand—
The arrogance his, the unhappiness hers;
Neither can get the upper hand,
The doing and the watching part
Outliving the torments of the heart.

Her children have outgrown it too,
But do they hear the twisting of the key,
Entail their ruined space in lieu
Of charity?

Love, orderer of dolls and towns,
Has Lilliputianized the scale of pain,
So the wide adult eye looks down,
Bereaved again

Of esperance, the childhood flush,
And has no passage into afternoons
But through diminished doors and hush
Of darkened rooms.

The shade of Thomas Hardy presides over both of these poems, in their anatomising of loss and pain, and in the formality which both controls and intensifies their attitudes. Irony runs through 'His and Hers' like a trace element, beginning with the usually-benign allocations of the title, and grief runs through 'Doll's House', beginning at the same point. Both have, in a sense the reverse of the usual one, economies of scale. The governance of the little household of each poem requires finesse in exposition, orderly disposition, succinctness. Donne's ambition to 'build in sonnets pretty rooms'[6] has its expression here too, though 'pretty' has a drier sense than any he intended.

When one thinks of classic pastoral, ancient or renaissance, it has a vertical and a horizontal axis. The vertical runs from the high affairs of state and the disposition of full-scale cities to the Arcadian revels of Theocritus, Milton or Pope: the horizontal runs between nymph and shepherd, between fabled and human beings. Porter, without being dutiful about it, uses similar axes. The man in 'His and Hers' has come from the 'glory' of the high boots to a box, the house from its earlier high estate to depletion; and the grim contest between man and woman lives on emblematically after their death. In 'Doll's House', the object of attention, as if it were a furnishing in a poem of George Herbert's, is a locus for allegiance and interrogation, at which the observer can gaze from a different dimension, but which speaks to him whatever his dimension may be.

Good poems both though they are, the second is noticeably the finer, partly because of the way in which scope of reference has been brought so exactly into narrow compass. For instance, the second stanza here gives us a modulation, in 'Love', of the 'charity' of the line before it: an allusion to Eros, builder of cities, and a reference to children's disposition of their little households: a recalling of *Gulliver's Travels*, so often in its abbreviated form a children's book, but in which Gulliver, in Lilliput, offers both benignity and naïveté: a 'wide adult eye' apt for alertness, hugeness, and vulnerability—Gulliver was, after all, to have been killed by the Lilliputians unless he fled: and the poignancy of 'Bereaved again// Of esperance', the first words perhaps echoing 'Alone again/ Naturally', and the second pair perhaps the unavailability of the language of unimpeded communication, certainly the unavailability of hope. It would be careless, too, not to make the point that the attraction of pastoral is usually bound up with its ability to lyricise the unattainable—in which lies its pang.

In one of Christ's parables,[7] the king says, 'Because you have been faithful in small things, I will put you in charge of great ones', and that is the way it works in pastoral; also the way it works with Porter. His apt reader needs to be able to read both fast and slowly—not once for the words and once for the music, but rapidly in order to get the architectonic and the logic, and slowly for the sake of the connotations which proliferate. Both of these ways of reading are ways of looking for exactness, fit between mind and word, fit between word and world. Of course, such 'fit' can take many forms: the guardee's uniform fits in one way, and the Bedouin's garb in another. To think about Porter's sense for 'fit', it is helpful to note some of his comments in an interview with Paul Kavanagh, published in 1985:[8]

> I noticed in your interview with Les Murray, that he states quite firmly that he is against the 'summing-up' notion of a poem. He feels the poem should simply end because, like a piece of land, it ends when it meets the sea, or when it meets something else. He says he likes to distribute the ends through the poem. That is a clever observation. I am less cunning than Les on the one hand, and less natural on the other. I do have a natural tendency to summing-up, but I have overcome the banal desire to encapsulate, or for the O. Henry surprise ending. I have often written poems where I knew the ending before I knew anything else. This could be a corrupt way of writing, but it is a possible way of coming to terms with just what the

imagination can do. If you know where the goal is, the next task is to discover the various channels by which you can approach the goal. It is almost impossible to describe in words, even though words are the medium of your work of art, the process by which the work evolves, is discovered, or is worked out. I only know that when I am writing a poem my feeling about it is that I am not going to be able to shrug off the responsibility of shaping it in some fashion or another.

. . . The art of poetry is a strange art, because it can be seamless, yet it can be taken apart like Meccano. If one recognizes that memorable speech—which is what Auden at one time called poetry—the talent for phrase-making, for providing the aphoristic nudge, is one part of poetry; while the other part of poetry is the opposite, is the inclusive, the weaving, the spellbinding, the making of landscapes, the putting together of all these, one sees what one is up against. One is always likely to be more one than the other, but sometimes poets are able to reconcile the two. Shakespeare did, clearly, and also someone like Pope, who was a great phrase-maker, but also extremely good at laying out whole terrains in which things can happen. Auden is a mid-way case, a poet with a natural gift for quite vulgar journalistic turns of phrase, yet at the same time, a concept of a larger territory in which the poem can live.

. . . In a way, poetry is like driving an electrical current through a high resistance coil, that becomes incandescent because of the resistance in the material to the progress of the current. This cannot be used literally about the imagination, but poetry might be thought of as what happens when the force of the imagination, which may be no more than a disquiet, itself perhaps produced by a kink in language, is driven through the material of words, so you get an incandescence. That incandescence then has to be frozen into shape, so that it can be reincandescent the next time someone reads it.

. . . Nobody, even the poet who says he is writing from direct experience, really knows what sponsors his notions. I like to think that the mind is so infinitely complex a machine for recording, remembering, inventing, that no one explanation of how it works will ever satisfy. It is an intriguing system, an infinitely complex lens on to what happens.

Those last remarks are a salutary warning against too-ready an assumption that one knows the 'how' of what is happening in poetry, in its reception as well as in its making, and they are certainly a rebuke to any reductivity; but the earlier remarks alert us to much that is habitual in Porter's work. Murray's phrase about distributing the ends through the poem surely does apply, in one fashion, to Porter. 'His and Hers', admittedly, is a comparatively early poem, and may in Porter's mind precede the 'overcoming' of 'the banal desire to encapsulate', but the fact that the poem could, not without loss but still with integrity, have ended after the first stanza or the second rather than with the third, represents something which is still lively in Porter's writing. Phrase-making and landscaping, to use his terms in the second quotation, go on having it out with each other in much later poems. Who, without knowing the poems concerned, could tell with certainty which of these formulations begins, middles, or ends a poem?

> I fear that the gods are awake after sleeping and mean to be
> noticed.
>
> We are not put on earth to be happy but to ensure
> The effective production of Daihatsu Hatchbacks—
> not even the Japanese could want to live in Japan.
>
> In the code of invisible painting
> trees are pushing colour to its limit,
> a green which parses understanding.
>
> His was the head which notes had chosen
> To move within—
> What gods and scientists had frozen
> Melted in him.[9]

The image of the incandescent coil, itself a bit bedazzling, is well-chosen, when we think not only of poetry at large but of Porter's activities. It picks up his polar or tensional habit of mind (as do, incidentally, all of the short quotations from him which I have just given); it implies both resistance and receptivity; it is small-scaled but laden with power. One can see all of these attributes on display, and the coil-analogy vindicated, in many of Porter's essays in lexical splendour—in 'Cat's Fugue' (*LIACC*, *CP* 228), for instance, or in 'Woop Woop' (*PW* 5): but they are there too in a poem in a lesser key like 'Frogs Outside Barbischio' (*PW* 49):

How reassuring to listen to frogs once more
From stagnant water in an old brick cistern
Beside olive trees run wild and the unprogrammed
Flight of a butterfly over hot fields and terraces.
One grandfather frog stays on his stick to watch
A self-tormentor return to his book to trace
His anatomy of melancholy. He's in Italy
To surprise an old hopelessness known long before.

The cosmos of frogs inside its wet-walled fort
Warbles and cavorts in the all that there is.
Wise frog rejoinders have challenged that book:
Come down to our waters so pulsingly black
And lose all your stubble of fortune and truth.
Here's art inside art, incision and sign
Of the purposeless minute outlasting its span,
Of the gloat and the plop and the stick still afloat.

Anyone acquainted with Porter's work will know that the best that the word 'reassuring' can offer him is the reiteration of what he is assured of, and that is largely bad news; the Eumenides are always just around the corner, even in beloved Italy. The butterfly's flight is congenial in its liberty, but the other side of liberty's coin is randomness, which does not do much for our spirits. And whatever is in the poet's vista, what is in nature's is an absurdist figure, the annuller of local possibilities, who plies his book with Burton's assiduity, to come out with Burton's eloquently sad circular conclusion. The country, by now for Porter long also a country of the mind, holds 'an old hopelessness' which matches the poet's, and which could in its grim way nourish his if that should fail. What happens here (and in the second stanza at, say, 'cosmos', 'pulsingly black', 'incision', and 'purposeless') is exactly 'the force of the imagination . . . driven through the material of words', sometimes deflecting or expanding the words, sometimes redirected by the words, and with them creating that 'incandescence' of meaning of which Porter speaks. This is what might be called the tragic use of language, fated and free, even when it is put to comic purposes.

'Distributing the ends through the poem' can take many forms. Auden wrote an 'Ode to Terminus',[10] the god of borders: Porter is fascinated by the liminal. The poetry of frontiers, of interface and fence and shore, of gateways and harbours, is his natural métier. There is a correlative of this about which I will be speaking later—transposition and transmutation—and the gestures of the mind which deal with the one are like those which deal with the other; but the liminal is an essay in precision, which is what concerns us at the moment. Somebody said that Shakespeare liked iron because it was hard and feathers because they were soft: Porter likes both precisions and fluidities. 'The Poem to End Poems' (*PW* 37) itself ends, 'Thus while each word remains I will discover/ The source from which the magic rays are sent/ And publish it this once, the world at play,/ Our single sphere which purrs with measurement'. The 'single spheres' of his poems purr with measurement, with determination.

Sometimes this is obvious, as in the rhyming coupletry of poems which are implicitly homage to Pope: 'Pisa Oscura' (*TAO* 70), 'The Sanitized Sonnets' (*TLOE*, *CP* 141), '"Talking Shop" Tanka' (*TCOS*, *CP* 265), 'A Bunch of Fives' (*PW* 38). It is less obvious, until sought, in the many poems which turn on juxtaposition and sometimes contrariety, though occasionally Porter does the thing with almost pedagogical plainness, as in 'A Chagall Postcard' (*PW* 17):

> Is this the nature of all truth,
> The blazing cock, the bride aloof,
> The E-string cutting like a tooth,
>   The night that crows?
>
> The cock has seen the standing grain,
> The bride is shrouded by her train,
> The violin is strung with pain,
>   A cold wind blows.
>
> From earth to sky the cry ascends,
> What breaks will threaten where it mends,
> Proud lovers end as pallid friends,
>   These feed on those.

That third stanza makes express what has been hovering in the two that preceded it: figure on ground, cessation and incessancy. It is hard to think of a single well-known poem of Porter's in which that configuration is not in one way or another the norm.

Porter's is an elegant poetry, if by that word we mean not gloss and sheen, but something like what it means in mathematical or scientific contexts: shapely decision. The abundance of idea, image, and trope in his poems can divert attention from an essential austerity of mind which finds its expression in apt collocation—apt, even if often surprising. His is, after all, a judicious, almost a judicial, way of writing. It resembles the conclusions which are arrived at by judges in the English tradition, who have to proceed by precedent, law, and fact, and from this jungle of givens find a path to an outcome. The 'opinions' of the great judges show comprehensiveness, and a corresponding decisiveness. This is what we get from Porter, in his own style. The judges often write as if they want two things: seamlessness of presentation, and precision of cut. The two can be found together in many of Porter's poems—in 'A Chagall Postcard', for example, or in this, from 'The King of Limerick's Army' (*PTTC*, *CP* 180):

> We can face such concepts here, our army
> has its rations, the weeds, the jokes, the softening air;
> Down a black hole to tumble; well, the church
> was there before us and Father Fogarty understands—
> If it's a matter of years, the Lord has many,
> and a field or two of Limerick besides.
> Spacewards fan the radio waves, scientists
> tingle like small boys with tadpole jars,
>
> Ninety per cent of the Universe is dead
> but our king's army is in acorn time
> And blustering by the stream his powerless horde
> undoes tomorrow what is fixed today.

It takes someone of Porter's capacities as a fantasist to scorn adequately the parochial delusions portrayed in this poem, and that is much of its brunt. It takes a different kind of skill to manage, in this and the poem's other passages, a simulation of the mind in its lurches, while keeping a 'governor' on the run of the sentences, in that highly distinctive combination of run-on phrases and capitalised new moves—'Down', 'If'—a device recurring throughout the poem. What we are looking at here is more than stage directions for voices heard, though they are given too; we are overhearing a mind trying for explanation, and seeing one way of mocking its pretensions while honouring its aims. We know the good judge not when the clear case elicits lapidary judgement, but when tangled circumstances do not impede exact delineation at the end. That is elegance, and that is what we get from Porter.

Pascal distinguished between *esprit de géométrie* and *esprit de finesse*, favouring the second for certain purposes.[11] As a great mathematician, he was well placed to do this; it was not the carping of the ignorant. I think that this is the kind of thing that Porter is up to in most of his poetry: he is courting the *esprit de finesse*, the perfect pitch for the moral and aesthetic actual. He tries, and tries again: he sounds like Wittgenstein, like Yeats: he sounds like Socrates. The melancholy which spreads its thin coat over everything he does is like rust on armour, patina on copper domes—always noticeable, at some future stage disabling, but frequently a diversion from what is going on beneath it: a wield of wit, a hunger for delineation, a plea for the truth. The medieval moralist said that 'manners maketh man': Porter, realising that art and human styles alike depend on one's knowing where to draw the line, is engaged constantly in the search for exacting illumination. Suggestively, in reviewing Les Murray's *The Daylight Moon*, Porter salutes Murray's 'dramatic sense of moral discrimination', even though he has clear reservations about some of Murray's conclusions.[12] His phrase seems to me an excellent one for what is happening in much of his own work.

5 GENERALISATION

In Whitehead's triad, generalisation re-engages us with the world at large; newly instructed, we are well-placed to be gripped anew. The simplest thing to be said about Porter's poetry under this rubric is that it is capacious in imagination and address. What he favours is made clear on more than one occasion. Here are two versions of it:

> Purifying the language of the tribe is only a feasible idea after the event. Before a work of art is made, the materials lie scattered around the artist in their impure form. I can see no reason why the modern writer, buffeted by commercialism, bad journalism and clichéd speech, should not make from these ingredients a work of the purest imagination. There is always the example of Wallace Stevens . . . of all American writers after Henry James perhaps the most emblematic of the poet's good sense and power of survival. Everyone knows that Stevens was a successful businessman, but most forget that his art has a shining relevance to the world he lived in. He made no fuss about the state of America, but it is there in his poems. Most studies of the American Dream are shoddy beside *An Ordinary Evening in New Haven*. The unpolemical Stevens reminds us that it is earlier, not later than we think . . .[1]

> The greatest poets have the least need of the arcane. They can make the most shining of structures out of everyday material. What we see is transfigured, but it keeps the lineaments of its ordinariness. Thus the process of inventing a poem and the process of composing music are similarly concerned with thematic transformation. When George Herbert turns cottage brooms and kitchen tables into great religious poetry, he is doing the same thing that Beethoven does with the notes of the common scale. Poets can learn from composers how to be more poetical, and that is a lesson needed desperately today. The battles between fixed and open forms, between tradition and modernism, between the shock of recognition and the challenge of the new are, finally, unreal engagements. Each true poet has the hard

> task of quickening the dust of the real and his only resources are statement, contrast, development and counter-point—the same as those used by musicians. Prose can afford to argue, to be discursive and compendious. It is like the sounds of Nature: it has not been selected and winnowed for its immaculate shape. Poetry is a selection of words from the whole store available, as music is of natural noises. Thus the two arts are doing the same thing, but they often help each other only as isolated acts of kindness.[2]

When a poem of Porter's is over, we have more access than we had at the beginning to the patrimony of being. This is not because he has been giving an overview, though he often gives that too, but because he has been making re-entries into it, curiously, hungrily.

The manner may combine diffidence with competence. His admired Lichtenberg says, variously, 'Whenever he was required to use his reason he felt like someone who had always used his right hand but was now required to do something with his left', and, 'That is as natural to man as thinking or throwing snowballs'. Porter would respond spontaneously to both of these—as to the rueful humour of, 'Never before had a mind come to a more majestic halt'. Porter is one of those who instinctively talk about 'the world'. The world can be the principle of reality, the unbeatable antagonist, the arena of conflict, the sum of being, the co-conspirator with flesh and devil, *tout le monde*. In all cases it is some kind of interlocutor, some kind of intervener upon narcissism or solipsism.

Porter's poems are all ways of 'going out into the world', even while they allow for the fact that the world may conduct its affairs enigmatically or with hostility. The promptings for his moves may be either that, to quote Lichtenberg again, 'Every stick of sealing-wax reminded him of the faithlessness of man and of Adam's fall', or that 'We can see nothing whatever of the soul unless it is visible in the expression of the countenance; one might call the faces at a large assembly of people a history of the human soul written in a kind of Chinese ideograms'.[3] Looking out or looking in, he is not just staging or gauging manoeuvres, but establishing forms of mutual interpretation between the items on which his attention is focused, even when he is one of those items.

When Johnson, in his imitation of Juvenal in 'The Vanity of

Human Wishes',[4] begins, 'Let Observation, with extensive view,/ Survey mankind, from China to Peru', what we are offered is generalisation, particularisation, and most importantly the *spring* between them, a spring found in but not confined to 'observation', 'extensive', 'view', and 'survey'. That is the kind of thing that has got into Porter's poetical blood. His carefully-geared, dexterously unfolding sentences, judging but tolerating, are likely to offer locale, world, and span, with the emphasis between the three shifting from point to point in any given poem:

> The painted pale projectile moves and seems
> Hardly to move to those inside its tide
> Of twilight, as the close alliterative
> Miles are ticked off charts of cloud and ocean
> And still the pilgrim's face is held to sun
> And still his stomach turns. A comb of words
> Runs through his strands of thought continually,
> Reminding him of God and lunch, of dust
> At the end of the day discountenancing shoes,
> Of mango trees and winds in March. These props
> Are human, more so than the glittering sprays
> Of alcohol, the swimming pools of in-
> flight movies, or reminiscences of men
> Returning to the Gulf. Packed into fear
> With splendidly unlikely comrades, he
> Out-herods Herod in this great escape
> And relishes the massacre.
>
> 'The Rest on the Flight', *TAO* 12

The title derives from a standard *topos* in painting—the Holy Family's rest in their flight from Herod into Egypt. Porter may seem an improbable member of that company, but then the tale would not have been retold and repainted were it not supposed to trope for us all an element in the condition of mankind. Porter has said,[5] 'It's utterly unimportant to me whether Jesus Christ . . . lived. The point is, there is a Jesus Christ-shaped need somewhere and it was filled. It's one of those questions that comes up as people make alarming conclusions about how Time works.' He does not share that need, but he does share a need for vivid tales which can, as he relives the human agenda himself, be reinscribed according to the modes of his own imagination. At all events, my interest here is not in his being knowledgeable about a biblical fragment and an array of paintings, but in the movement between universality and particularity, palpable time and insinuated stasis, words our servants and words our masters.

Anyone who has travelled enough for banality to have set in will recognise in this poem the tediousness of life and the fear of death, but will also find in their combination an emblem of much of adult life, at least as it is apprehended in solitude. 'The painted pale projectile moves and seems/ Hardly to move to those inside its tide/ Of twilight . . ./ And still the pilgrim's face is held to sun/ And still his stomach turns . . ./ Packed into fear/ With splendidly unlikely comrades'— this is at once highly theatrical and, for some, a commonplace. The hubristic style of all commercial airlines, as if they were indeed the siblings of Pegasus or Mercury, not to speak of God, arcs around the discourse in which they characterise their mechanical venture. Porter's imagination is one which can both entertain and criticise this commercial imagination, and all those who consent to at least part of its tranquillising tale: 'the rest on the flight' is also a resting in flights of fancy, even though it calls for the palliation and the stimulus of 'another/ Miniature', and yet another. Running, as usual, all through this passage, as through the whole poem, is the busyness of deed; verbs pervade the verbal for him, almost as if it had no other office than to carry them, and they no other role than to dramatise speech as the prototypical element of restlessness.

When Porter deals with 'life' or 'the world', as he does with unabashed incessancy, he does not address them simply as 'things in general'; he comes to cases, and in those cases redivines the promising insights already available to him. Bruce Bennett calls a chapter in his book 'Fuel for the Dark', quoting from 'On This Day I Complete My Fortieth Year' (*TLOE*, *CP* 139), Porter's personal 'piling on fuel for the dark'; everything I know of Porter's activities in language points to the aptness of this citation. But we are not looking mainly at an ensemble of tallied items: we are looking at a stream of processes, of negotiations between existing perspectives and test cases. To use the term much in vogue some years ago, Porter's art is one of bricolage, of improvisatory initiative. The objects in front of the bricoleur, while perhaps 'shining', cannot hold attention for long unless he can deal them into the negotiating of time, the correlations of understanding, and the summoning of imagination.

Nature's own bricolage is metamorphosis, and poetry's is above all metaphor. Both are central to Porter's work, and it is on these that I want to concentrate for most of the remainder of this essay. Aristotle,[6] writing on rhetoric, says that 'We should give our language a "foreign air"; for men admire what is remote, and that which excites admiration is pleasant': and, on poetry, 'The greatest thing by far is to be a master of metaphor. It is the one thing that cannot be learnt from others; and it is also a sign of genius, since a good metaphor implies an intuitive perception of the similarity in dissimilars'. Porter has not always had good words for Aristotle, but Aristotle would have to give good words for Porter's usual way of proceeding: metaphor is his staple. So, in 'Returning' (*ES*, *CP* 280),

> Nobody feels well after his fortieth birthday
> But the convalescence is touched by glory
> So that history's truculent deeds of hate
> Are lived through in dreams, the story
> Followed to the investigator's hut, pain seen
> Through a window on its knees, late
> Help lost over marram dunes or never
> Felt at the deliverance on a screen.
>
> Marvellous means of escaping time and time's
> Chosen people: sleep on the knowledge of
> God's monsters! The school of love and crimes
> Is open every night and the sedentary
> Heir of men of action dashes off
> A sonnet before execution. What you see
> Is coiled in an uneventful past, rough
> Justice of the body's failures, a commentary.
>
> Yet never daring enough, even those hours
> When the timid rule of truth relents
> And every written word is without sense
> As in some ultimate avant-gardish shape—
> The apostle of plain dullness has powers
> Of arrest and will use them; nightmares
> Are prized categories too, a southern rape
> Modelled in blood but with a classic tense.

It is time to recompose the face
Into a serious map, the children now
Envied creatures across a room, the case
Being settled for the present. Home is
The veteran of the adjectival run,
His images intact. He has learned how
To live another day and wakes, ringed
By the golden wallpaper of the sun.

One of the 'returnings' anatomised and celebrated is that from the world of dreams to the world of the day, the passage through the horn gates and the ivory gates. Porter's poetry is steeped in dreams. They fascinate him, no doubt, for many reasons, which is unsurprising, since it is hard to imagine any artistic culture in which dreams did not fascinate, but one of their uses is to give him metaphors for metamorphosis. Dreams, at least as deployed by Porter, are the very embodiment of mutation. A few poems, such as 'The Three Dreams' (*APF*, *CP* 78), 'Dreamtime' (*LIACC*, *CP* 223), and 'Dream Restaurant' (*PTTC*, *CP* 165), signal explicitly what is to come, but there are hundreds of references to the visionary, illusive and elusive contents and character of dream in Porter's books. He has an educated interest in traditional and modern 'readings' of dreams. As with so many other of his resources, those readings nourish his writing, but he is not held to any consistency in what he makes of the dream-conditions which he records or invents. This, after all, is the man who said, 'I hate all forms of confirmation, because I like the imagination to be continuously in suspension, and never to come down on the side of anything'.[7] He is not likely to become the captive of any orthodoxy of interpretation of that phenomenon which offers him most mobility.

Anyone who is as absorbed by the phenomenon of dream as Porter, and as given to noting the wildness of much human behaviour during the waking hours, is likely to see the borders between the two conditions as fairly arbitrary. 'Fled is that music:—Do I wake or sleep?',[8] 'He awoke and found it truth'.[9] These romantic dicta might be things overheard in the background of many of Porter's poems. The life he portrays is full of transgressions, not only of the ethical and the socially tolerable, but of the codes whereby it is commonly modelled or appraised at all. Swift's notion that life is neither aggrandised to tragedy, nor released into comedy, but is a travesty, finds its echoes at many points in Porter's writing. In fact Porter, like Swift, and Rochester, often sees humanity as self-travestying, as inherently dysfunctional, and as self-bewildering. Given poems of his may, like the collected aphorisms of Lichtenberg or La Rochefoucauld or Chamfort, constitute improvised systems for the intellectual and emotional management of an otherwise intolerable turbulence. They cosmofy the chaos for a while.

And so, in 'Returning', a poem whose title might be the slogan of polymorphs, things melt and solidify constantly. If you look at Lake Michigan in winter, you may see thousands of frozen waves—fluency in captivity. 'Returning' looks like that. Formally, it has enough constancy to model structure, and enough instability to model change: it is a poem of cycle. Many of the things within it answer to or echo one another: 'touched by glory' and 'the golden wallpaper of the sun'; 'the story', 'every written word', 'a commentary', 'a serious map'; 'convalescence', 'deliverance', 'means of escaping', 'the case being settled'; 'the school of love and crimes', 'the apostle of plain dullness', 'the veteran of the adjectival run'; and so forth. As usual with Porter, there is at least enough meaning to be going on with. And that is what he does: he goes on with it. Shelley, in the *Defence of Poetry*,[10] says of poets, 'Their language is vitally metaphorical; that is, it marks the before unapprehended relations of things and perpetuates their apprehension'. I take the last phrase in a strong sense, in applying it to Porter. The apprehending is intensified and mobilised; it is the seeing as well as the seen that engages him.

There is not a line in the poem which is quite at rest—certainly not the last line of any stanza, including the final one, with its operatic overtones of the ring of fire, and its mythic and Yeatsian overtones of the golden apples of the sun. Almost any poem of Porter's is likely to announce or insinuate an element of problem, and this one does in its first line, and then play physician to its own distress, as this one does too. But just as all doctors, in the long run, lose all their patients, Porter remains clear-eyed about the degree to which he can and cannot command the situation definitively. After all, the 'apprehension' of the 'relations of things', even while the apprehending is heightened, is not the same as comprehension; some of the world of meaning goes on running through the fingers. 'Home is/ The veteran of the adjectival run,/ His images intact' is a genial claim, but the waking poet does not settle as Stevenson's sailor or hunter does in his home; there is too much sailing, too much hunting, still to be done.

Seamus Heaney is fond of quoting Yeats's saying, 'The end of art is peace',[11] and of teasing out its meanings. There is something of great importance at stake here, for anyone thinking about the arts of poetry. I doubt whether Porter would agree fully with Heaney's claim. It seems to me that poetry is systematically dissatisfying, has salt or alcohol in it, and leaves us both pleased and athirst when a poem is concluded. Certainly, Porter's poetry reads like that to me. I often think that that poetry reads as if its author is at once the insider as to understanding and the outsider as to audience. I do not mean by this either that it is arcane by policy or that Porter's instincts are hieratic. The references are as accessible as they need to be for anyone who takes poetry as seriously as a scientist takes science, and if anything Porter is anti-hieratic. I mean rather that there are some kinds of realisation which make the person who undergoes them feel disconcerted and isolated.

Porter, asked whether he writes for an ideal reader, says, 'Yes, I write for an ideal reader: myself',[12] and although it is said cheerfully it has, too, a plaintive element. To go the metaphorical way, to be in what might be called a 'state of metaphor', is for him to be at grips with understandings that disconcert even when they define. 'Pain seen/ Through a window on its knees': 'time's/ Chosen people': 'a southern rape/ Modelled in blood but with a classic tense': the hold that such things have on the imagination is not attributable solely to violence or to vividness. It has to do besides with intelligence's folding experience into its own care, conscious that it is called to illuminate, but unable to guarantee adequate illumination. If the end of this art is peace, it will not be a self-congratulating peace.

'Truth', says Malebranche, 'is distant, it is not palpable, and it is not a good which we feel impelled to love. It calls, therefore, for the closest attention';[13] Lichtenberg says, 'We do not think good metaphors are anything very important, but I think a good metaphor is something even the police should keep an eye on.'[14] These are two forms of attentiveness, and they coincide in Porter's 'The Rival Poet' (*TCOB* 45):

His work shines through the spaces of your thought
So crystalline for being words in stars
And phrased as if our blood itself could speak.

The world he sees is cuneiform for God
But he is free of moral adjuncts, dining
With names and beauties in allusive dachas.

He has the dangerous beard of a Tuscan
Innovator early dead: you see
Him strut like Tolstoy on his own estate,

Each peasant pregnant by him; he has flowers
At his fingertips which praise the land
While you are keyboard-bound and colourless.

Call him a louche Longinus and he'll smile
And dandle a child upon his knee and scan
The family snapshots for sublimity.

The first to make a picture picture its
True self in diagram, he lets his mind
Recapture infantile omnipotence.

His pre-shrunk paradise is metaphor,
A crossword puzzle running out of clues
Till even death-beds start to stink of skill.

It is your vanity to be his Other,
Erecting trellises of words to keep
The world suspended in the winds of time.

Two politicians of the printed page,
You both are talking to a vacant moon.
The tongueless stones have said it all before.

Somebody has been through this neck of the woods before, and his name is John Wallace Stevens Ashbery, or perhaps John Robert Browning-Donne. But neither of these shadowy individuals has taken out a lien on the territory's possibilities, and Porter makes some of them his own. The fact is that metaphor, whether or not a 'pre-shrunk paradise', is amenable to almost any degree of fidelity or infidelity to 'all that is the case', is

Prospero's onus and Ariel's flair, is that perilous diomorph, Apollo blent with Dionysus; and the other side of the same fact is that beyond metaphor we have nothing. The very word 'literal' is a metaphor: the very word 'metaphor' is a metaphor: expand the hermeneutical circle as we may, a circle it remains. The poet, like his batman the critic, may at any moment regard this as blessing or as curse, and Porter is able, in this tonally subtle poem, to modulate between these two attitudes. Looking at the burden of 'His work shines through the spaces of your thought/ So crystalline for being words in stars', or of 'The first to make a picture picture its/ True self in diagram', one is compelled not just by the vigour but by the valour of what is being done; this is what they used to call '*inventio*'—the finding of the sources of imagination, as people tried to find the source of the Nile. And looking at the burden of 'Till even death-beds start to stink of skill', or of 'Erecting trellises of words to keep/ The world suspended in the winds of time', one is haunted by senses at once of beauty and of fragility, of cables of gossamer. I think that Porter is astonishing in the full-heartedness with which he has staked his lifelong venture as a poet on the centrality of metaphor. He is not bewitched by metaphor, nor flirtatious with it: he simply lives in it, is indebted to it, is quizzical of it, and bears it, as we do all these things in our own body. Metaphor is to Porter what the sea is to the Homer of the *Odyssey*: the milieu in which all that can happen for good or ill will happen.

There should be a slogan, 'making strange is making known': and effectively that is what the whole stress on *ostrannie* amounts to—as the same point does in Aristotle. The trivialisations of this are easy to imagine, easy to recall: *dérèglement de sens* is all very well when synaesthesia is at issue, as it is fairly often in Porter, but it has little interest if we are thinking about the searching possibilities of the intelligence. A more responsible gesture, and one very consonant with Porter's practice, is the one rehearsed in Heaney's poem, 'Making Strange': 'I found myself driving the stranger/ through my own country, adept/ at dialect, reciting my pride/ in all that I knew, that began to make strange/ at that same recitation.'[15] This is taking hold, and it is letting go, and we know that the touch is right in both moves. Heaney's reference to his 'own country', a terrain in which he has found himself variously at home and an alien, is a reminder of

Porter's complex dealings with what might be called his 'own countries'. Be they Australia, England, Italy, or elsewhere on the map, or be they realms of the imagination for which he alone can provide the visas, they are constantly under revision—metaphor, after all, is incorrigibly revisionary.

This scanning attribute of metaphor, its refusal to let being hold priority over becoming, is often rehearsed in Porter's titles. They are often taking him into or out of scenes, predicaments, contexts: making camp anywhere, he is already thinking about breaking camp. 'Walking Home on St Cecilia's Day' (*OBTB*, *CP* 13), 'The Great Poet Comes Here in Winter' (*PAAM*, *CP* 62), 'Encounter in Antioch' (*PAAM*, *CP* 73), 'Inspector Christopher Smart Calls' (*PAAM*, *CP* 75), 'Steps on the Way' (*APF*, *CP* 91), 'The Last of England' (*TLOE*, *CP* 119), 'On the Train between Wellington and Shrewsbury' (*PTTC*, *CP* 171), 'Tending towards the Condition' (*PTTC*, *CP* 226), 'Night Crossing' (*LIACC*, *CP* 206), 'Down Cemetery Road' (*LIACC*, *CP* 226), 'Exit, Pursued by a Bear' (*LIACC*, *CP* 233), 'To Venetia's Shores Have Come' (*TCOS*, *CP* 262), and so on: they signpost territories for which new signposts will be needed by the time Porter has finished with them.

The exercise is carried through, sometimes, in one kind or another of *paysage moralisé*. Auden gave an authoritative version of this in 'In Praise of Limestone'.[16] In doing so, he moved beyond the comparative rigidity of the allegorical mode to a way of imagining in which both tenor and vehicle were 'estranged'. Word and world were both played with more generously, and were both taken more seriously, than Auden's 'placing' of man the worldly creature would seem to promise. Perhaps, for Porter, this poem, Auden's own favourite, had a significant influence. In any case, his own imagination has had a disciplined hauntedness when place turned metaphor has been at issue:

> But the stickiness underfoot was disquieting,
> Perhaps the land was Avernus, with those
> Bamboo raggednesses above the fence and the pale smell
> Of warm tar on the air. Through fur of sugar-grass
> He saw the river and all remote existence
> Sculling across the darkened tide. What blew in his face
> Were words, those he would speak in love and those
> Which fattened on betrayal. The words for death

Were still unknown and yet he knew they sought him
On the street. A wind of big mothers mixing drinks
Caught him suddenly with laughter. What if he got
To the wharf, what if the ferry with a Lady's Name
Were there? He sang, in case, 'Goodnight, deceiving world!'

'Landscape with Orpheus', *ES*, *CP* 309

There are, of course, bland natures
back at home, whose only earnest
is a theory that your expedition
means to straighten out the land with names,
to fit a grid of the accounting gods
on plastic otherness. You'll have to give
a hostage to them—do it now.
Here by Disappointment Bluff, gaze
across the Vale of Sixty to Uncanniness;
beyond the rock-strewn creek
is an escarpment called Incalculable
and the fields are more elided than Elysian—
mark this tree the point of going back
and set it on the map—Porter's Retreat—
the place at which all further progress
ceases to have consequence.

'Porter's Retreat', *PW* 50

Kafka wrote of the point of no return—'that is the point that must be reached'. 'Must' in Kafka always has a unique inflection, but some of its combination of doom and summons is an element in Porter's sensibility—as is some of Kafka's sense of ironies buried in the language. Kafka also wrote, 'The word "*sein*" signifies in German both things: to be, and to belong to Him',[17] not a sentence which could give Porter much heart; but 'Disappointment Bluff', 'the Vale of Sixty to Uncanniness' and 'more elided than Elysian' are language's ambiguous gifts to being. John Olsen, writing of Donald Friend, refers to a 'discipline of inferred knowledge',[18] and this is powerfully at work when Porter goes the metaphorical way. The 'knowledge' is partly cultural and partly organic, as the ageing body and spirit speak of life's common courses; and it is 'inferred' both in the sense of something calculated to and in the sense of something delivered over to the reader. When Porter 'places' himself metaphorically, as in both of these poems, he is as we all are in literal places: surrounded and thus controlled, but mobile and thus free. Metaphor is his preferred mode for the exploration of this universal predicament.

What rarely ensues is a sense that ultimately the impedimenta of imagination can be shed and a heavenly exit or entrance be had; for him, to be alive is to be in a state of metaphor. 'What blew in his face/ Were words, those he would speak in love and those/ Which fattened on betrayal': it is the leitmotiv of all the Porterian music. Wittgenstein writes, 'We speak of understanding a sentence in the sense in which it can be replaced by another which says the same; but also in the sense in which it cannot be replaced by any other'.[19] Porter's *paysages moralisés* have the lucidity of the first kind of sentence and the luminosity of the second kind.

In the end, all of the dream-work and all of the metaphor-work is at the service of metamorphosis, which is both Porter's compulsion and his game. If dreams fascinate him, it is because they are both such shape-shifters themselves, and such apt emblems for much else in life. Dreams remind us of films and paintings, which recapitulate drama, which enacts personal volatility, which echoes the mutation of seasons, which instances the blend of stability and mutability in the whole world, which Shakespeare sometimes thought a dream, and so on, *da capo*. Significantly, praising Shakespeare, he remarks that 'his talent for producing metamorphoses was greater than Ovid's',[20] and that 'The model for art remains that of metamorphosis: imaginations, great events, are all transformation scenes. It is no surprise to learn that Shakespeare's favourite poet was Ovid.'[21]

Metamorphosis dwells amongst the inglorious as well as the glorious. The early 'Metamorphosis' (*OBTB*, *CP* 10) begins and ends thus:

> This new Daks suit, greeny-brown,
> Oyster-coloured buttons, single vent, tapered
> Trousers, no waistcoat, hairy tweed—my own:
> A suit to show responsibility, to show
> Return to life—easily got for two pounds down
> Paid off in six months—the first stage in the change.
> I am the only image I can force upon the town . . .

I look to the wild mirror at the bar—
A beautiful girl smiles beside me—she's real
And her regret is real. If only I had a car,
If only—my stately self cringes, renders down;
As in a werewolf film I'm horrible, far
Below the collar—my fingers crack, my tyrant suit
Chokes me as it hugs me in its fire.

This is not prentice work, but it lacks some of the imaginative suppleness which later became Porter's hallmark. The suit of Nessus at the end has come off someone else's rack, and throughout there is agitation rather than the agony which later came to show itself. But the conceit in the title is representative, and is to have many successors; and the later poems would not have their often moving character if they too were not prepared to include ignominies, the humiliations which life deals us whether or not we advert to them.

Ovid and Ovid's many debtors have beautified transformation, but many of the changes which they describe literally brutalise those who suffer them. Time, the ultimate transmuter, is not famous for beautifying human beings physically, and is an uncertain bet when the moral or the spiritual is at issue. Porter, unquenchably curious about temporalities, has few illusions about what the years, the decades, or the centuries may bring. But then, as with sexuality in 'How Important Is Sex?' (*ES*, *CP* 277) or in 'Student Canteen' (*FF* 33), the temporal is not to be wished away, even though it has to mean the temporary. The first poem in *Once Bitten, Twice Bitten* is 'A Giant Refreshed' (*OBTB*, *CP* 3), the last in *The Chair of Babel*, 'Le Jardin Suspendu' (*TCOB* 68): they are equally concerned with the expectation and the inevitability of change. As Nature, at the origins of modern science, was said to be 'put to the question', an expression taken from the torture-chamber, so in Porter's work he and his fellows are constantly being put to the question by the transmuting hand of time:

. . . Do you think
They knew they lived in an Age of Eloquence
Who were arguing with stars through fiery glass,
Who took perfection of the artist's form
For granted, working instead on how steam fell

And rose in tubes and how the ugly module
Of just seventy years might blaze a golden rose
For Demos? And in the Age of Epigram
When an out-of-kilter cummerbund
Or a wrong caesura in hexameters
Was most of what was worrying in art
Some brutal primitive was marketing
A colossal apparatus raising myth
To high symphonic shouting . . .

'Le Jardin Suspendu', *TCOB* 68

Intransigent metamorphosis in the self, in society, or in Nature can be taken for mere spectacle, as it often is in Ovid. It is hardly ever like that in Porter. Frequently, an important part of the poetry is a self-preserving wit in the midst of nature's or culture's shifts and excesses. Chamfort writes,

> The physical world seems to be the work of a powerful and good being who has been forced to abandon to a wicked being the execution of part of his plan. But the moral world seems to be the product of a devil who has lost his mind . . . Often the mind is to the heart merely what the chateau library is to the lord of the manor . . . I would be happy to say of metaphysicians what Scaliger said of the Basques: 'They are said to understand each other but I do not believe it.' . . . They asked Madame de Rochefort whether she wanted to know the future. 'No', she said, 'it's too much like the past'.[22]

The first of these sounds like the beginning of 'The Cats of Campagnatico' (*FF* 41), the second like 'Reflection on my own Name' (*OBTB*, *CP* 14), the third like 'The Return of Inspector Christopher Smart' (*APF*, *CP* 93), the fourth like 'The Irradiated Poem' (*TCOB* 28). All of these poems, like Chamfort's sayings, mirror our hurts; but each of them, like every mirror, recomposes those hurts, and offers us the momentary illusion of permanency.

When we think about metamorphosis in Porter, it is helpful to linger a little with his 'gods'. God-talk in Ovid is a way of flirting with the absolute and the relative at the same time. No conventional philosopher would look twice at Ovid for any kind of systematic insight; but he might take from Ovid a metaphysical frisson which had to do not with divinity, but with our being so parti-coloured, so pied, in our awareness of some absoluteness in life while we are unremittingly conscious of change.

Certainly, the 'gods', as they have been troped over more than two thousand years in Western literature, have often stood in for that doubleness of attitude or intuition. I think here of two books, Jean Seznec's *The Survival of the Pagan Gods*,[23] and Leonard Barkan's *The Gods Made Flesh*.[24] Seznec traces, up to the Renaissance, the ways in which the figures of the gods were continually reappropriated in order to emblematise or betoken meaning, in cultures ranging from 300 BC to that of Shakespeare's England. My guess is that in all this Ovid was recurrently apropos not only as a site to be quarried, but as the symbolic patronal figure for the notion of what might be called deployable divinity. In any case, Seznec's account is a suggestive one when one thinks of those figured presences, sometimes mocked, frequently ironised, but peculiarly tenacious, which crop up across the range of Porter's poetry. 'Gods they had tri'd of every shape and size', Dryden writes contemptuously of an erring people;[25] in Porter's poetry, gods of many shapes and sizes crystallise the apprehensions of sensibility and the formulations of experience. He may absolutely disbelieve in them, but they surely do bring him metaphors for poetry.

Porter is on many counts a thoroughly different poet from W. S. Merwin, and yet when I read them both I find that they frequently remind me of each other. This is not because of, say, their common indebtedness to dream as motif and condition, or to their uncommonly formidable air of prevailing personality, or to the fact that they both pasture on distress. It is rather because of their shared sense that the absolute mutates with a fatal constancy, and that what poetry may often do is bring this home to the appropriate reader—the appropriate one being the one who is prepared to be changed in order to take this in. If I were asked to cite poems of Porter's which would be relevant to this claim, they would include, 'Dorothy Osborne in the Country' (*APF*, *CP* 82), 'Steps on the Way (*APF*, *CP* 91), 'The Sadness of the Creatures' (*TLOE*, *CP* 136), 'The Great Cow Journeys On' (*PTTC*, *CP* 181), 'The Picture of Nobody' (*TCOS*, *CP* 238), 'What I Have Written I Have Written' (*ES*, *CP* 286), and many others.

In Merwin's case many poems could be cited, and so could

some of the remarkable prose pieces collected in *The Miner's Pale Children*.[26] One of these, 'Humble Beginnings', goes as follows:

> When he had learned how to kill his brother with a rock he learned how to use a rock to begin stairs. For both of which secrets he thanked the rock.
>
> He considered the rock further. It had always been there keeping secret what it could do. It had never so much as hinted at what it had already done. Now it was keeping all its other secrets. He fell on his knees facing it and touched it with his forehead, his eyes, his nose, his lips, his tongue, his ears.
>
> He thought the rock had created him. He thought that.

Another, 'The Travel Figment', begins,

> It is one of the many unacknowledged gifts of Hermes, the God of Delusion, to those images imprisoned behind their own eyes, men. He had to deal with them. He got to know them, in the manner of dealers. He watched them dragging their shadows, in ignorance—those boneless black banners that were all that was left of their wings. He watched their blood groping through their brains like the ghost of an amputated limb, looking for the sea. He was moved—not by pity, of course, to which he is eternally immune, but by an impulse to elicit from mortals a fuller admiration of his own command of distance, his grasp of the elements, his ability to transmit himself instantly from world to world. He was moved by vanity, of which he is one of the lavish princes. And by his own love of idleness—he thought he might make his missions easier, by making his activity more apparently comprehensible.
>
> He gave men the idea of travel . . .

Porter could not have written either of those pieces, but the temper and sensibility at work within them could find a place in many of Porter's poems. Even the titles of a number of things in *The Miner's Pale Children* could have been Porter's: 'The Basilica of the Scales', 'Postcards from the Maginot Line', 'The Weight of Sleep', 'The Medal of Disapproval', 'The Dark Sower', 'The Second Person', 'The Permanent Collection', 'The Fragments'. Over all of these there brood the fickle, peremptory, and inexorable presences of the gods.

Barkan's *The Gods Made Flesh* demonstrates the way in which metamorphosis has played a decisive role in the West's apprehension and reconstruction of antiquity. Barkan writes, 'I discovered in my first adult reading of Ovid's *Metamorphoses* the origins of all that most fascinated me in Renaissance literature and art: not the sources of the stories (splendid as they are) but the genesis of a world-view and an aesthetic'.[27] Porter had his own world-view and aesthetic from elsewhere, but the Ovidian ingenuity, copiousness, and constant inconstancy are things deeply sympathetic to his own evolving project. There is in Porter's work no such thing as a poem which offers, purely, a vista of retrospect or prospect. The historical gaze is a gaze into entities and energies which offer to transform the attention they require. Some examples of this would be 'The Historians Call Up Pain' (*OBTB*, *CP* 30), 'Vienna' (*PAAM*, *CP* 49), 'Homage to Gaetano Donizetti' (*PAAM*, *CP* 61), 'Europe: An Ode' (*TLOE*, *CP* 127), 'Frogs at Lago di Bolsena' (*LIACC*, *CP* 212), 'Where We Came In' (*FF* 29), 'Susannah and the Elders' (*TAO* 46).

Eva Hoffman's memoir, *Lost in Translation: Life in a New Language*,[28] divided as it is into 'Paradise', 'Exile', and 'The New World', might have been written to give a title and a typology to Porter's work. When aged forty-eight, in a piece called 'Locked out of Paradise', Porter wrote,

> Some years ago I wrote a Lowellian poem of family history entitled 'Five Generations'. It is not among my best works, but rereading it today, I am more concerned by its cavalier indifference to fact than by its rather spavined style. The facts may not be absolutely wrong, but they are certainly doctored to make the poem more interesting. That has always been my way when writing poems—I fictionalize them. This springs less from basic dishonesty than from my only hazy sense of what actually happened in any situation . . . when I use my own life as material for poems I tend to go . . . into a world of moral and aesthetic aphasia. Until I have remade the facts I don't know what they are.[29]

Elsewhere, writing on literal translation, as distinct from Hoffman's kind, he says,

> The Bible may well be the greatest example of poetry as metamorphosis in existence. We print the ecstatic Hebrew poetry of the Old Testament as prose, and the wonderfully rich language which the

> committee who produced the Authorised Version of 1611 wrote in has gone on to influence every subsequent generation of English poets all the way to Dylan Thomas and, in our own decade, Peter Redgrove and Galway Kinnell. There is even a form of this sea-change which could be described as 'creative misunderstanding'. Out of a false reading of the original an artist of powerful imagination invents a new order of sensibility in his own language. The power of the Bible is so great that the many metamorphoses of its texts, though framing in turn the characters of whole nations, may still be excused the charge of being true misunderstandings.[30]

Reading this, one is conscious not only of its aptness as cultural commentary, but of its symbolic appropriateness to his own work. In the same article, he provides 'pigeonholes' for the ways of 'remaking' within translation. These he calls, 'Uncompromising Scholarly', 'Aesthetic Scholarly', 'Recreative Scholarly', 'Exuberant Scholarly or Over the Top', 'Literal', 'Imitation', and 'Ecumenical Slovenly'. With the probable exception of the last, these could serve as metaphors for Porter's own exercises in the 'translation' of world to word. There are constant habits of mind, favourites for attention, and items inspiring revulsion, but the modes in which the world is alluded to, pried at, celebrated, or decried, are unpredictable—as his most recent book, *The Chair of Babel*, continues to show. In his literary and musical journalism, he is always on the alert to the possibilities of transformation. I note only a couple of examples. Writing in 1976 under the title, 'Miracles of Metamorphosis: Gustav Mahler', he says, 'Mahler's art is all transformation—change, metamorphosis, not just of themes and their harmonies, but of programmatic ideas associated with the themes, of emotions and circumstances . . . Mahler performed miracles of metamorphosis with the 'banality' of his obsessive material'.[31] And in 1978, reviewing a collection of Auden's writing,[32] he quotes from the poem, 'Happy New Year':

> On a lorry the centre of a gaping crowd
> A man was eating a hedgehog whole;
> Presently he rose and said very loud
> 'The colon we know is the seat of the soul
> Keep the colon clean by conscious control.'
> In the middle distance a titled whore
> Was distributing trusses to the ruptured poor,

and asks, 'Am I being far-fetched in seeing the origins of "The Shield of Achilles" in the stanza just quoted?' He is not being far-fetched, I would say; but only someone fascinated by the possibilities of transformation is going to mention the link as well as notice it.

I want to draw this handling of metamorphosis in Porter to a conclusion by quoting from a writer whose concern is not with his work, but who put suggestively what I would like to claim about his writing. The quotation is from A. D. Nuttall's book *The Stoic in Love*.[33]

> Shakespeare produced in his auditors and readers a profound sense of the indefinite richness of reality . . . It is a curious experience to turn from Shakespeare's dramatic writing to a popular Renaissance handbook of style such as the *De Copia* of Erasmus . . . One finds an immediate kinship with Shakespeare, and, at the same time, an immense difference. Erasmus shows his readers how to say 'Your letter gave me great pleasure' in 148 different ways. The general effect is of a rigid proliferation of mere forms. Yet it is similarly characteristic of Shakespeare (unlike, say, Racine, who is the master of the single, perfectly encapsulating, closed phrase) to offer a series of alternative phrases. In Erasmus the variations offered are so to speak variations only. There is no ontological implication. Their sole interest lies in the ingenious manner in which they differ from one another as forms. In Shakespeare, we sense instead that these are alternative 'essays', shots at a target, nets cast out to catch some ulterior reality which is at once elusive and excitingly important. Thus at once, at the level of mere style, although we find convention and selection, the mode of that convention is one which *obtrudes its own incompleteness*, and so quickens our sense of transformal possibilities.

It will be clear by now why I have quoted this passage. If it is true that, as Lichtenberg says,[34] 'what has to be done in the world in Shakespeare's way Shakespeare has for the most part done', it is also true that some of the tilt of his intelligence is to be found elsewhere, and certainly in Porter. Those alternative essays, shots and nets are the kind of thing that is his staple; and 'the obtrusion of its own incompleteness', not for its own self-afflicting sake, but to 'quicken our sense of transformal possibilities', is one of the major attractions of his poetry.

6 Three Poems—Some Pointers

'ALCESTIS AND THE POET'

'Alcestis and the Poet' (*ES*, *CP* 292) occurs, in the *Collected Poems*, between 'Garden of Earthly Delights' (*ES*, *CP* 291) and 'Addio Senza Rancor' (*ES*, *CP* 294). The first of these ends, 'Pavilions of self-love/ Adorn the grass: and here you must build': the second, 'You are nowhere/ in the evening light: what I see instead/ are two white presences, playing with life,/ smiling and letting it go without reproach.' The lodgement of 'Alcestis and the Poet' just there may be accidental, but those conclusions suggest important terms for its development. Pavilions of self-love can be monuments to narcissism and to selfishness, inviting reproach, but pavilions of self-hatred would be nothing to rejoice in, either; these alternative nightmares can haunt Porter's poetry at many points. On the other hand, the poet can farewell the two women portrayed in a photograph album—'one was to die/ at forty-one and the other at forty-four'—as those who knew how to live and how to die. 'Alcestis and the Poet' takes some of its moving force from the interplay between two such impulses—to acknowledge ineptitudes in the conduct of life, and to salute imperfect boldness in an imperfect world.

Euripides's story of the woman who offers her own life so that her husband Admetus may live has had various redactions, one of them by Rilke, whose poem, in Stephen Mitchell's translation, ends, 'One more time he saw/ the girl's face, for just a moment, turning toward him/ with a smile that was as radiant as a hope/ and almost was a promise: to return/ from out of the abyss of death, grown fully,/ to him, who was still alive—// At that, he flung his hands before his own face, as he knelt there,/ in order to see nothing but that smile.'[1] As usual with Rilke, there is virtue in that 'almost'. I do not know whether Rilke's formidably beautiful poem influenced Porter's, but it would be remarkable if someone so long an admirer of Browning, and so influenced by him, had not been touched in the writing of 'Alcestis and the Poet' by Browning's 'Balaustion's Adventure',[2] his reworking of Euripides's story. John Pettigrew[3] says that 'the portraits of Alcestis and Admetus, it is generally felt, are affected by points of resemblance to Mrs Browning and the poet himself'—something to the point when one reads Porter's poem as one of his versions of his living and his wife's dying.

Browning, who called his poem 'A May Day Amusement', and who brings Alcestis back from the dead, has his narrator Balaustion record, of the Golden Age which the two would supposedly inaugurate together, that she could not learn 'That ever one faint particle came true,/ With both alive to bring it to effect:/ Such is the envy Gods still bear mankind!'[4] In Porter, almost all givings are attended by misgivings, and had he needed licence to see things so, he could have found it in Browning, whose principal figures usually generate more perturbation than they quell. He never needs that licence, of course. At one point, Browning writes, 'And, as some long last moan/ Of a minor suddenly is propped beneath/ By note which, new-struck, turns the wail, that was,/ In a wonder and a triumph, so/ Began Alcestis',[5] but this modulating of a minor into a major chord by altering one note in it is not the sort of thing one looks to see in Porter. Still, if one wants to 'place' his poem imaginatively, one could do much worse than triangulate it from Rilke and Browning.

Other poems which I find illuminating of 'Alcestis and the Poet' are 'The Shining God and the God of Correction' (*APF*, *CP* 87), 'Night Crossing' (*LIACC*, *CP* 206), and 'Dreamtime' (*LIACC*, *CP* 223), each of them a dream-poem in which, as in the medieval poems tagged in that way, waking guilts and griefs are given a rendition. 'Alcestis and the Poet' is not a dream-poem in quite that sense, but it is charged with the peremptoriness of fear and of reproach to be found in those earlier poems.

The title, as so often, is more than a label. Alcestis's address to her Admetus is also one to her maker, her deployer; it is as if Euripides, Browning, or Rilke had found himself, in disconcertment, to be Pygmalion. It is important to remember the extent to which Porter has perceived himself as, has made himself into, 'the poet'. Some poets do their making with their left hand, as it were; not so Porter, for whom the whole thing has been pursued with a passion not to be disguised by the necessity to make a living otherwise, or by a conviviality of spirit in social or pedagogical circumstances. He has a plentiful capacity for self-mockery even in that role, but it is native to him, and he is almost unimaginable without it. For the creator to be turned upon, or at least to be called to account, by his creation, is a curious imaginative experience: none the less so for its being fairly frequent in Porter's writing. The effect is all the more impressive in that there is a bonding between the two figures. There are sentences in the poem (for instance the second, the ninth, the fourteenth, the seventeenth to the twenty-second) which could as appropriately come from the poet's mouth as from Alcestis's. This is relevant, because in all versions of the Alcestis story some dwelling on the two-in-one-flesh motif is central; the death of the self-sacrificing wife is to be the death of the man left behind, as well as his life.

That fact about the situation is signalled in the words, 'I/ Have spent my life in your service', something given a different accent in the later, 'We are ready to do each other service'. If the 'service' is partly a Muse's in the second case, it is personal in both cases; but the quality of service is very various in range. One feature of Porter's poetry is that any element of experience, particularly elements of the imagination or of personal relationships, is open to being redefined. This is not true of some poets, including some great ones, such as Horace or Wyatt. But each of Porter's poems is a kind of mutability canto, in which the given is put into the cauldron of consciousness to be tried and to be refashioned. That is what is happening in this poem to the character of 'service'. There is a mutuality of service between the poet and the one who is, suggestively, his 'subject', and the process of the poem reopens the question of their relationship. It often seems to me that much is gained when reading a poem if it is addressed as though pristine—as though someone had just invented the art. A poem like this one lends itself to the notion, inasmuch as it invites a revision of one's understanding not only of its topic, but of the terms on which its topic could be addressed. It is in fact not only about Alcestis and the poet but about Alcestis and poetry.

To say that it is also a poem about death is to state the obvious: to say that it is 'about' death—as in Donne's saying of the mind's quest for truth, that it 'about must, and about must go'[6]—is less obvious but equally true. Porter beats the bounds of meaning, and never more energetically than when meaning seems recalcitrant and evanescent. To think of this poem as simply a way of 'getting over' the death of his wife would be as imaginatively unalert as it would be impertinent; we are looking instead at an engagement in the relationship between memory, metaphor, and metamorphosis.

One needs help when engaging in a venture of this sort, and amongst Porter's resources are the existing traditional story, a sense of solidarity with earlier artists (Shakespeare, Chatterton, Stevie Smith), and voices, variously attached and unattached, to assure him that he is not as alone in his activities as loneliness and grief are prone to make us all suppose. His mind, like Donne's, is that of one both grateful for access to large contexts, but fundamentally provisional, a darter's, a freebooter's. This poem is the product of such a mind, and it also bears the marks of the influence of similar minds, Browning's perhaps, but Shakespeare's certainly. Again and again one can hear the Shakespearean cadences, can recapitulate the Shakespearean gestures: '. . . heard the chorus of self/ Desert its lord to swell your tragedy . . . we glow among our feelings/ And are pointless . . . The soft arrival counts . . . we are all for darkness . . . The rest/ Is timing . . . The wind urges the trees to sigh/ For us: it is not a small thing to die . . . Casting words upon the page'. Several of these are invocations or revocations of celebrated lines from Shakespeare, and all of them catch his voice. In a poem in which heard voices animate both the argument and the emotions, the presence of the master's master, the imagination's equivalent of God, is one of its principal resources. And just as in the most carefully pitched monologues of Shakespeare one sentence falls upon the next, and falls away from the previous one, in 'Alcestis and the Poet' Porter offers process achieving formulation, definition hungering for dissolution. 'So,/ As young Shakespearean gestures, we glow among our feelings/ And are pointless. Then, as the shades of madness/ Intervene, we become important': it is done with the instincts of an iconoclastic lapidary.

There is another way of looking at the poem, and that is to think of Brodsky's friend's magical cloth, which takes out some elements and leaves in others. If we were to find various qualifiers left in, adjectives and verbs, they would be very suggestive. I single out some of them: 'little', 'wounded', 'early', 'useless', 'young', 'pointless', 'important', 'soft', 'tongue-tied', 'inexplicable'; whatever this is, it is a cadencing of the emotions. It is impossible to understand Porter's poetry unless one grasps at the outset that the incidental intellectuality is at the service of intense and pervasive feeling. In some poetry, the resources of the intellect go to establish a kind of monument to meaning in the midst of the flux of the emotions: this was, perhaps, the principal grandeur of the eighteenth-century poets. In other poetry, those same resources, while treasured and commanding tribute, are reinvested in the flow of feeling—what might be called Wordsworth's debt to Milton becoming an endowment for Freud. Porter's originality consists in part in his having a gratitude for classic definition while remaining free for romantic enthusiasms, more especially the dark enthusiasms. We might say that for the Romantics, articulation without aura was not really articulation; if so, Porter is deeply invested in the Romantic pursuit.

Once again, the figure of Auden is an instructive one to invoke. His didactic style when writing prose always ran to speaking of polar alternatives—Eros and Agape, Prospero and Ariel, Nature and Grace, the tough and the tender—and all this amounts on the face of it to an analytical aloofness. But surely those commentators are right who discern in Auden, whatever the restraints laid on the heart, a lavishness of the imagination. Auden's poem 'A Summer Night' has lines which gave Gavin Ewart the title for one of his own poems—'The Lion griefs loped from the shade/ And on our knees their muzzles laid,/ And Death put down his book'. Ewart's poem ends by referring to 'Auden's wonderful hybrid rose that crossed the comic and the tragic'.[7] Porter has learned how to make the 'hybrid rose' not only of the comic and the tragic, but of the Classical and the Romantic. But the 'cross', as in the rose, is not annulment but liberation; one is given not less feeling but feeling intensified even when scrutinised. To pair 'Are you there/ Where each new disappointment makes you think that life/ Is geared to reparation?' with 'But looking back I see only the disappointed man/ Casting words upon the page. Was it for this/ I stepped out upon the stairs of death obediently?' is to observe appraisal put to the test of passion.

All this is a way of talking about the presence of the poet to his poem, or in his poem. Someone whose interest was, much more than mine, in stressing distinction between presence and poem, between *parole* and *langue*, could find much to exercise him in reading 'Alcestis and the Poet'. It is clear that the poem collates precedents and confects an outcome: Euripides, Shakespeare, Chatterton, Stevie Smith—their writings constellate whether or not their presences endure. Among those 'someones' would be Alcestis herself, the scrutiniser and rebuker of the poet's longing to identify himself with his poetry. The voices singing in German are foreign, however beautiful: Chatterton, 'marvellous boy' though he may have been, was a death-embracing boy nonetheless: Stevie Smith's highlighting 'Death . . . in the small print', while dashing, was alien in its stylishness: the 'young Shakespearean gestures' made early, or the 'little gestures' that 'The creatures that we are' make late are little more than gesticulant. The paradox that precisely where we seek to instantiate meaning we may most readily be perceived as gratuitous and thus meaningless is the real nightmare that hovers around the poem for most of its course.

'Alcestis and the Poet' concedes that the Porter/Admetus who is being addressed can never mount any adequate case in his own defence, given that it is precisely for dexterity that he is being put in the dock. The sentence, at first sight puzzling, 'The king sits in his kitchen,/ Not certain if the world knows of her sacrifice,/ Though don't his cats despise him?' is not attended to fully if we think of some combination of domestic nursery-rhyme and Porter the lover-of-cats-in-poetry. We have to think as well of the man's being dethroned from the estate of art by the woman's giving of her life for his life—a perspective complicated but not lessened by whatever Jannice Porter's suicide may have done to or for Porter. When thinking of these matters, it is useful to remember what he has said when speaking of 'An Exequy', (*TCOS*, *CP* 246) another poem derived in some sense from the experience of his wife's death:

> There are risks of tone, and in fact a lot of people have objected to things in that poem. They have noticed the simple form and the reminiscences of Bishop King, but they have not noticed the departures from Bishop King, so the poem is actually saying something rather different from what they think it is saying. If I told them what I thought I was saying, they might be shocked, because it would interfere with their notion of what is proper for an exequy.
>
> The poem does state a number of odd things. I refer to Piero, who

> is Piero di Cosimo, not Piero della Francesca, and the picture I have in mind is the famous painting of the stranded monster, with Andromeda tied to a rock, a rather camp looking Perseus about to despatch the monster, and people standing all around. Now, I was identifying my wife, not with Andromeda, but with the monster. When I think of the picture, my sympathy is entirely with the beautiful, but rather grotesque monster, not with the noble warrior Perseus, or the simpering heroine, Andromeda. You need a formal poem to contain the tone of something like that. So I used the advantages of the simple form, the rhyming octosyllabics, having the cadence of King's 'Exequy' in mind, and tried, because the poem is a genuine lamentation, to move in and out of the orthodoxy, to show how these things are more complex than, for all its Caroline elaboration, the version you get in King's poem.[8]

We are looking, in other words, not at the transcription of experience but at its transformation, and the same may be said at almost every point about 'Alcestis and the Poet'.

'To show how these things are more complex' might be Porter's watchword in most of his poetry. Emerson said that 'Literature is the effort of man to indemnify himself for the wrongs of his condition'.[9] Wallace Stevens, that 'Poetry is an effort of a dissatisfied man to find satisfaction through words, occasionally of the dissatisfied thinker to find satisfaction through his emotions'.[10] True, no doubt: but the effort can be tactically various, tonally complex. Alcestis says to her poet, 'will you/ find the inexplicable, the out-of-reach-of art/ Intensity you mourn for outside Hades?', and the poem may be said to be a series of 'casting words upon the page', where 'casting' has overtones of casting spells, casting lots, casting down, making casts at, casting the die. The modes are likely to be mixed—with, for instance, the ironic domesticity of 'giving up a good position in the queue' paired with an echo of Iras's, 'we are for the dark', near Cleopatra's end. Ambitious poetry is always on the verge of acknowledging that 'whereof we cannot speak, thereof we must be silent': one way of continuing the quest while acknowledging its near-futility is to go on shifting the pitch, raising or lowering the mode. That has long been one of Porter's resources. Sometimes, simplicity will do all that can be done: 'the soft arrival counts . . . The rest is timing . . . it is not a small thing to die'. Even then, though, a context of investigation and reappraisal has been established. The sublunary ones will not find 'the things we've sacrificed' pristine and waxing, either on the moon or here below, but the best consolation may be, for the disappointed, to look for those 'sites' about which 'great art' may come. Nothing could be more typical, though, than for Porter's poem to end with a question about the whole affair, a question plangent, partly resentful, imaginatively haunting, and unanswerable.

'CROSSING THE TIBER ISLAND'

This fierce and moving poem (*TAO* 69) is a more overt, less self-qualifying piece of writing than is often the case with Porter. Few of his poems remind one of Lowell, but this is one of them—the Lowell of 'The Dead in Europe'[11]— 'After the planes unloaded, we fell down . . . Here hugger-mugger in the jellied fire'—or of 'Beyond the Alps'[12]—'Now Paris, our black classic breaking up/ like killer kings on an Etruscan cup'. It is a poem of outrage mediated through some of the bleakest of historical ironies, and grappling with some of history's densities as if, this time, they do more than offer matter for sombre reflection, and will not let him alone. Its combination of religion, statecraft, violence, insanity and cruelty is almost unique in Porter's writing. If Porter has not produced even the most original of guidebooks to Rome—his transforming energies are as strong here as anywhere else—the poem will not work as it should unless one is alert to some of the context upon which he is working his poetical will.

The savage jest of the first line impinges as something in which Shakespeare has once more had a hand. 'As flies to wanton boys are we to the gods,/ They kill us for their sport':[13] the entertainment offered by later circuses has as its provenance the Theatre of Cruelty which developed in late republican and in imperial Rome. Around the Circus and the Colosseum that theatre organised itself. John Pearson's *Arena*[14] gives a good deal of information and insight which are relevant here. He reproduces Piranesi's etching of the Colosseum, which shows how the residua of Rome can loom hauntingly in the imagination—as they are doing in Porter's poem. The dense, intricate, and brutally efficient building towers over the tiny figures inspecting it; as in Piranesi's *Carceri*, man has become little more than a speck before the inhuman entity reared by his own ingenuity. Of the ethos of the killing games, Pearson notes:

> Static and self-absorbed, Rome was becoming a theatre in which all that mattered was that the show continued . . . The Emperor showed himself to his subjects, a godlike creature crowning an immense mass spectacle from the imperial box at the Circus or the arena . . . He was above the crowd, above the arguments and petty bickerings of state. His was a superhuman presence, and the claims of his divinity merely gave form to what he was in popular imagination. It was this image of the Emperor as God that began to loom above the great mass spectacles of Rome . . .
>
> The drawing-power of the arena was clearly enormous. It was not simply for the rabble. All groups and classes found themselves drawn in, so that the games could come to form the central unifying mass activity of this enormous empire . . . In his Easter pilgrimage and annual service at the Colosseum the Pope becomes as nowhere else the visible successor of the Caesars . . .
>
> Finally, Caligula's cruelty overflowed; all Rome became his arena and the terror started. Citizens were arbitrarily arrested, tortured, put to death. Men of good family might be branded for no reason, thrown to the beasts, and even, according to one account, sawn in half. There seemed no reason for these persecutions except the Emperor's madness . . .
>
> The Colosseum was at the centre of an effective system of empire-wide mass anaesthesia . . . it helped to turn the exuberant Republic into the deferential empire of Augustus, and finally into the static, stagnant universal state, far more concerned with watching the arena than with repelling the barbarians at its gates.[15]

Quoting this suggests something of importance about much of Porter's poetry: its engagement with history. He knows a great deal and feels the weight of things. Some of the most impressive of his earlier poems have had to do with the decline and fall of the British Empire, as this impinges upon a watchful figure come from further away than any barbarian came to witness the end of the Roman Empire. When Porter deals with the contemporary, one knows that he is more than abreast of the quotidian: in London or Brighton, one senses not only plurality but pressure. The same thing often happens when he recapitulates more remote times and places: 'Soliloquy at Potsdam' (*1962/3*, *CP* 41), 'Encounter in Antioch (*PAAM*, *CP* 73), 'At Lake Massaciuccoli (*ES*, *CP* 306)—nobody goes to these places like

Porter, but there is a distinctive feeling to each of them when he goes there.

It would be an instructive exercise to assemble a volume of Porter's poems which were, in one sense or another, 'Roman'. These would include the Martial versions, poems like the present one and 'Roman Incident' (*TCOS*, *CP* 269), and the *Narcissus* and *Mars* sequences. What would then be seen, I think, would be a reciprocal pressure between history and myth on the one hand, and Porter's recreative reception of them on the other. What one gets habitually from Porter is something analogous with what happens in his prized Mozart: a motif elaborated, refined, deepened, contradicted, and in one way or another resolved. He is not a poet for solo instruments or voices, as it were: too much happens, too fast and too intricately, for his mind's music to be attended to at only one level at any given time. Consider, for instance, some of these features of 'Crossing the Tiber Island'; the first sentence gives us at least the following things:

A releasing of many of the implications of what is there in the quotations above from Pearson. Overtones are as important as what is said expressly; when Caesar quelled an incipient mutiny by addressing his paraded troops as 'civilians' rather than as 'soldiers', the contemptuous rebuke worked not by designation but by implication. 'God's Circus Maximus' is freighted with the implications of Porter's deeply ambiguous attitude towards 'God' or 'god', a figure knowable or imaginable in the end only through the various configurations and responses of creatures—performances which include impressive ventures in creativity, but which also include travesties of the true, the beautiful, and the good. Porter's quarrel with God is as intransigent as Job's, and he does not expect any answer out of any whirlwind. At all events, to offer God as the imperial observer (and presider, and builder, and maintainer) at the killing field of the Games is to unleash hostilities to the divine and to the mastering which will have their intensification and elaboration throughout the poem.

A bringing the game up to God, or at least to belief. The psalmist says, 'Even the sparrow finds a home,/ and the swallow a nest/ in which she puts her young—/ Your altars, O Lord of hosts,/ my king and my God!'(*Ps.* 84) Theology is of no interest to Porter; but Western civilization has until quite recently configured itself as believing in presiding divine presences which were incidentally (as in ancient Rome) or by definition (as in

Christendom) benign, and at best providential. But there is no future for *this* fledgling sparrow, no home in Rome, nor recourse to any altar, nothing to be had from 'king and God'. The issue—the ultimate issue in theodicy, the ubiquity of mortal suffering—is recouched a few lines on in 'Are not two sparrows sold for a farthing?.' This is Christ's question when challenging his followers to belief in a loving providence—something addressed tartly in the later, 'to ease Christ's vegetarian pain'. We miss much in Porter's poetry unless we notice the extent to which it is a mustering and a deployment of challenge. We are accustomed to seeing such a demeanour in, say, Neruda or Brecht, but in Porter's case it is as if he has reversed Charles Peguy's claim that everything begins in *mystique* and ends in *politique*:[16] Porter's distastes are political, but his revulsions are metaphysical. Rail though he may in conversation at the callousness or blockishness of politicians, his quarrel is Job's quarrel. Yeats's 'We make out of the quarrel with others, rhetoric, but of the quarrel with ourselves, poetry'[17] is no doubt as true of Porter as it is of Yeats; but he has a quarrel beyond 'the quarrel with ourselves', and it is the one which has always been the reliable progenitor of tragedy.

A linguistic mastery. The poem as a whole is unusually rough-cast, as if there were as little designing as possible to its utterance. Porter's mimesis, though, is never just a replication of life's accidental spontaneity, and the lines of this sentence are woven together with a number of touches which indicate command. 'Fledgling sparrow slides' offers the diminutive '-ling', a half-heard rhyme between 'fledge' and 'sledge', the 'slides' of small human creatures, and the awkwardness of utterance which looks for such smoothing out as will dispose of the sparrow itself. 'Miraculously' chimes with 'dynastically', the first implying the uncovenanted and thus uplifting, the second those illusorily reared establishments which we remember principally by their fall. The 'God' who should be doing better is leagued with the Pan of 'panic', the very paradigm of irrationality and self-absorption. The 'pavement' and the 'gutter', both inventions of the Roman engineers, define the fleeting terms of existence of this tiny being. Dwelling for a little on these six lines, one is reminded that, whether or not Porter always gets his lines right, there is very little coasting in them; words work hard for him, or they are not allowed into the poem.

A first movement in the process of the whole poem. What I

have been mentioning sets the terms for the operation of the poem from first to last. Porter likes cycles, rondures, thematic confirmations. Like many another poet, he does not really take to unfolding stories, to seriality and the linear imperative. Donne thought that decisive attraction would make his 'circle just',[18] and we often have the same effect in Porter. The poem ends, 'The sparrow in the dust/ knows neither Pontifex nor Aesculapius/ but twitches on its bed of wings/ terrorizing Heaven and whichever/ deity could meet its dying eye.' Much has happened in the twenty-nine lines between the end of the first sentence and the beginning of this one, all of it going to confirm the force of both, but the two sentences could in fact have gone together to make a poem in their own right. In Porter's poems, much novelty may present itself, and much penetration, in their unfolding: but he has an imaginative imperium of his own, which ordinarily shows itself within the first few lines.

It becomes clear, at least by the end of the second sentence, that the Tiber Island is a synecdoche for Rome itself, as Rome is for the world. Between them, 'creature' and 'gravity' do the job, the first encompassing created things as well as this pathetic being, and the second emphasising the gravitas on which the republican and then the imperial culture congratulated itself. Porter, 'crossing the Tiber Island', is retracing a path much more ancient than, say, Oxford's avenue for the Western cattle-drovers. The island is a site of myths and of religions, of duels with mortality. It was once sacred to Aesculapius, god of healing: it used to be a hospital 200 years before Christ, and it is still a hospital today.

All of this history simmers in Porter's consciousness. The island was once thought of not as an island but as *the* island, as Rome perceived itself not as a city but as The City. Porter's poem, with a strenuousness rarely found in his writing, delivers the island and the city as crystallisations of human amplitudes, human trivialities, human disasters. That 'bath night fronts a deathless Pantheon' is not only an incongruity of scales; it is a reminder of the religious transmutations of that unique building, and of the confrontation between mortality's dirt and detritus, and something which is at its best a metaphor for immortality. That Bernini's columns can be seen as the faecal outcome of a papal fascination with the succulent and the infantile is not something whose relevance is confined to Rome, to the past, or to Catholicism. That the Isola Tiberina is a natural springboard to the good eating of Trastevere—'the cows ten times your weight, the little/ prawns inside their carapace, and lettuce . . .' says something about our appetites, our passions. It also evokes subtly the milieu of the 'Games' once more, and the massive hunting shows mounted in the Colosseum.

In a curious way, Porter has made Rome his own imaginatively, more even than London. When, as in 'Santa Cecilia in Trastevere' (*FF* 44), located only a generous stone's throw away from Isola Tiberina, he writes, 'Her thunder/ is the face of beasts, apotheosis of/ a natural sweetness, death at the end/ sounding a miracle, tempering transcendence', we know that he is exhibiting a *pietas* towards this figure with whom he has had much earlier poetical dealings, but we also know that he has effectively stalked into Rome as Dürer or Montaigne or Goethe did, and taken the whole affair as imaginative spoils.

He is helped in doing this by its being 'other' to a degree that London—or Brisbane—could never be. Marianne Moore called Ireland 'the greenest place I've never seen'.[19] Randall Jarrell said that he believed that his favourite country was German.[20] In either case what was at issue was not just geography, and not just language, but the lingo for terrain, the topography of a dialect. Porter reverts to Rome, and more generally to Italy, as one for whom their singularities can never stale. This does not mean that he will accept the account given of either of those locales as authoritative; his scepticism of all authority tells against that. The congeners of Aeneas might have been content to 'cup . . . blood for dreams to drink', the inheritors of Peter the Rock resort to 'stones of faith outstaring nerves'; none of it washes very well with Porter.

One reason it will not do so is because of the centrality in his consciousness of consciousness itself, that element which can at any time practise critique—or run riot. 'Each human mind is Rome ruled by a mad/ and metaphorical Emperor, or Pope' occurs precisely in the middle of the poem, and in effect institutes a new stanza, another room in the imagination. That 'your history is just as marvellous as Rome's' is by no means all good news, given what many of those Roman marvels comprised, but the mind as marvel has always been one of Porter's preoccupations, whether he has framed it in terms of dreams, memories, guilts, or works of genius. His poem is a work of modernity not only in its contemporaneous trappings—the notorious Roman traffic, the hotel's air-conditioning—and not only in its scepticism, but also in its fascination with our capacity for ingenuity, even when that ingenuity is perverse. Pearson, reproducing Breughel's *The Tower of Babel*, describes it as 'a Colosseum gone to seed'.[21] That is a good description of Porter's sense of our minds on a bad day. Perhaps we all get the Rome we hope for or fear. The Rome Porter has got in 'Crossing the Tiber Island' is one whose geography and history and architecture are coarse-grained versions of the sinister city which he takes us all to inhabit.

'TALKING TO THE LIZARDS'

At the end of the French edition of Marguerite Yourcenar's *Memoirs of Hadrian*, there is a note, 'Rediscovered in a volume of Flaubert's letters, much read and heavily underlined by me around 1927, the unforgettable sentence: "With the gods no more, and Christ not yet, there was, between Cicero and Marcus Aurelius, a unique moment when there was man alone." A large part of my life was going to be spent trying to define, then to portray, this man alone yet connected to everything.'[22] The note evokes many reflections, and one of them is that 'man alone' is on several counts appropriate for the Cicero of Porter's poem (*PW* 59).

Not on every count: nearly 800 of his letters have survived. He was gossipy, volatile and vain: he was both ambitious and a vacillator, and incessantly preoccupied with the figure he cut. And yet he was certainly 'man alone' in important ways. Leader, for a while, of the Republic, he was to see it go down under the depredations of Caesar, Antony, and their allies; defeated politician, he went into formal exile once, forbidden to come nearer than 500 miles to Italy, and into retreat several times; and he was always glad, as in the present poem, to take to one of his country houses and survey the scene from there.

These are some of the elements in the life upon which Porter draws in 'Talking to the Lizards' (*PW* 59). As usual, his original is transformed. '. . . mere/ Anarchy, the state falling apart in kindness and service' is an echo of Yeats's 'The Second Coming';[23] the gentleman of letters and leisure at a distance from Rome has been crossed with the twentieth-century counterpart who was also disconcerted at violence's harm to the life of justice and harmony. 'The city is there/ At the end of all roads; may it forget me/ If I forget it' echoes the psalmist's imprecation upon himself[24] if he should forget Jerusalem—though 'The City' could do worse things to Cicero than forget him, and did, at Antony's hands, at the end. The 'dust . . . from planets as lucky/ As Caesar' draws on the astrology available to Cicero, but also on the knowledge of cosmic debris, 'stardust', which falls in tons on the Earth each day. The 'redundant lyrics' of the insects give us a couple of jokes—one Cicero's, in that 'redundant' is a favourite word of his, and one Porter's, in the play on the musical term, 'redundant interval'. Somewhere, too, behind 'I saw a face open which/ Had been a Roman's, a hard politician's whose love was on sale,/ Whose words were like nightmares where lampposts become/ Portable gibbets' there seems to be an echo of the end of Plutarch's Life of Cicero: 'On Antony's orders they cut off his head and the hands . . . to be placed above the ships' rams on the rostra, a sight for the Romans to shudder at, since they thought that they saw not Cicero's face but an image of Antony's soul'.[25]

Such reflections belong in night-pieces. By now, it will be clear that much of this book has been about what might be called Porter-by-night—about upheaval, solicitations, translation by dream. The bulk of 'Talking to the Lizards' is clearly operating in that country. What we have in the poem is a celebration of the mind's long combers, and a registration of storms at sea. Chesterton called the legions 'the walking walls of Rome';[26] there is a legionary drive to some of this poem, but also a shock when the drive meets intractable resistance. The first effect goes like this: 'Nobody should live all his time here, away from the dust/ Of delivery vehicles, away from/ The knowledge that news is arriving which turns men's complexions to/ Maps, tells them that labour's relentless as life,/ With queues of fine spirits sold up for breaking'. The second goes like this: 'My books are my truffles,/ I dig them in dreams'. In fact, some of Cicero's most celebrated speeches traffic in the memory, or the anticipation, of nightmare—of social dissonance in the place of harmony, of frenzy usurping all energy. A great one for mapping our rectitudes, this Latin Voltaire or Hobbes is also uniquely eloquent in denouncing their violation. Porter's dramatic monologue incorporates both elements of his spirit.

'What can't be cured must be endured'. Lamentation may not

be overcome by reason, but may be heard differently at daybreak. In 'First Things First',[27] Auden is visited one night by an awareness of love-speech and blessedness, by contrast with the usual 'smirking devil . . ./ Predicting a world where every sacred location/ Is a sand-buried site all cultured Texans do,/ Misinformed and thoroughly fleeced by their guides,/ And gentle hearts are extinct like Hegelian Bishops'. In the morning, he does not know how seriously to take the good intuitions, though he is glad that the storm that awakened him has put 'So many cubic metres the more in my cistern/ Against a leonine summer'; the outcome is salutary, even if not all issues are resolved. In 'Talking to the Lizards', the morning settles a tormented imagination, but one element in it is palpably false as argument. Whatever Cicero, as politician, could plausibly say of himself, it is not, 'We make music when we/ Act, almost as beautiful as real music,/ A patterned disinterest like the movement of limbs across marble.' 'A patterned disinterest' is not an accurate portrayal of Cicero's performance at any stage, even when he was mouthing the language of idealism.

That, though, does not mean that Porter has made a false move at the end of the poem. One needs to keep in mind three things when reading the poem as a whole, and the last ten lines in particular. The first of these is its investment in wordplay: the second is the bestiary tradition: and the third is the poetry of retirement. As to wordplay, the poem is a production of *homo ludens*. The notable amount of alliteration, the flicker back and forth between one sense and another of key words—'master of *morals*', 'country-*wise*, country-slow', 'Had been a *Roman's*',—and the syllabic structure throughout are part of the evidence for this. So is the swing between a Latinate and a more conventionally English sense to a number of words: 'complexions' as facial surface or as a deeper make-up: 'punctuation' as the swans' inscription on the surface of the water or as their point-to-point precision: 'convecting' as the bearing heat or bringing together, as against Cicero's use of '*convector*' as 'he who goes with me, a fellow-passenger': 'convulsed' as twitched, or as the Ciceronian reproach for the disarraying of the state. Running through the poem, in a minor key but audibly, has been the trace of a sense that all this has been a kind of 'ludic rehearsal', a playing over the surface of things. If 'the Innocent Age has slunk to its cave,/ The Epoch of Turbulence waits at the rim', this is no joking matter, but our wits are kept about us partly by the deployal of wit.

Then there are the beasts, great and small—the creatures. The swans, the insects and the hogs are minor touches: but the lizards, like the cats in a number of Porter's poems, can be regarded as being in touch with reality partly in their very otherness, and in their not being subject to second thoughts, as well as in their having 'No path to prepare'. Montaigne was irritated with Cicero for his not fronting questions as the Frenchman liked, for his havering—as haverer he could certainly be; but the Montaigne who wondered whether he was playing with his cat or his cat with him would have understood about the lizards. Sometimes, in poetry, small or great creatures comport themselves attractively just because they know what to do—as if they are life's experts, with man the only amateur. I think that this is part of the attraction of the lizards in the present poem. Cicero had so much work to do—almost like the Mozart of Porter's favourite quotation from him, who was fatigued less by composing than by not composing—so many levels at which to operate, so much watching over the shoulder to engage in, such personal insecurity that he had to go on scooping up all the applause he could get from anyone. By contrast, the lizards proceed, as Plato would have said, 'musically'—vitally, according to proportion, and in attunement with their world. Having nothing to say, they could still be eloquent in music; in the end, eloquence was the best that Cicero had to offer the world.

It may be that the most interesting thing about the poem is its sharing in the long tradition of the poetry of retirement or exile. The various moods displayed—geniality, world-weariness, low spirits, self-possession—are like those to be found in poetry written all the way from Cicero's time to our own. (Cicero himself wrote poetry, which adds an extra note to 'Talking to the Lizards'.) Wyatt, hustled out of London to cool his heels in the country, gives a catalogue of all those things that reconciled him to his lot:

> I cannot wrest the law to fill the coffer,
> With innocent blood to feed myself fat,
> And do most hurt where most help I offer,
> I am not he that can allow the state
> Of high Caesar, and damn Cato to die,
> That with his death did 'scape out of the gate
> From Caesar's hands, if Livy do not lie,

And would not live where liberty was lost,
So did his heart the common weal apply.
I am not he, such eloquence to boast
To make the crow in singing as the swan,
Nor call the lion of coward beasts the most,
That cannot take a mouse as the cat can;
And he that dieth of hunger of the gold,
Call him Alexander, and say that Pan
Passeth Apollo in music manifold . . .[28]

In poetical inscriptions a man is not upon oath, and we are not bound to believe that this was the whole story about Wyatt. Still, the sense that life in the country has it all over life in the city, or at least at court, was older than Wyatt, older than Horace. Plenty of the republican myth in Rome appealed to the figure of the honest countryman, collaborating with nature, exhibiting natural virtues, safe from the rampancy of ambition. It was always possible for Cicero to think of himself as being the guardian of the ancient values, of being himself *rus in urbe*, the country in the city. On the other hand, finding in the City the concentration of all the political power which he so valued in life, he could scarcely rest easy in the country. At least as he is portrayed in 'Talking to the Lizards', he exists at the nexus of many antitheses. Some of these are: day/night; city/country; animals/men; men/gods; the state/the selves; 'letters'/ deeds; harmony/turbulence; limbs on marble/ words on the page; morals/ songs. Yeats said that the good poet should be so simple that he has no faults but sins[29]—a saying farcical if taken literally, but suggestive if one is thinking of a certain kind of purity of being, a coherence betwen all the features of one's life. In that spirit, the conclusion of 'Talking to the Lizards' offers a hankering which is haunting—and all the more so in its invocation of music.

The imperial court came into being only after Cicero's death, but he could see it coming. 'Imperator' was the title conferred by the acclamation of troops in the field on their victorious general, and Cicero had seen too much of that not to understand what would come to be under Caesar or the Caesareans. A guarded trafficker in the gods, he would have agreed with the view that the faults and the opportunities are less in our stars than in our selves. In the end, 'Talking to the Lizards' is about what most things in Porter's work are about: the travails of consciousness, the consolations of art, and the calling to remain watchful to the last.

NOTES

INTRODUCTION

1 Randall Jarrell, 'The One Who Was Different', *The Complete Poems of Randall Jarrell* (Farrar, Straus and Giroux, New York, 1969), 316.
2 Christopher Smart, 'Jubilate Agno', *The Norton Anthology of Poetry* revised edition, eds A. W. Allison et al. (Norton, New York, 1975), 517.
3 Blaise Pascal, quoted in Jean Mesnard, *Pascal: His Life and Works*, trans. G. S. Fraser (Harvill Press, London, 1952), 199.
4 Donald E. Westlake, *Good Behaviour* (Mysterious Press, New York, 1985), introductory note.
5 'Three Poems for Music', *APF*, *CP* 101.
6 W. H. Auden, 'The Cave of Making', *Collected Poems*, ed. Edward Mendelson (Faber and Faber, London, 1976), 521.
7 Ursula K. Le Guin, *The Word For World Is Forest* (Berkeley Publishing Corp., New York, 1972).
8 Abram Tertz, *A Voice from the Chorus*, trans. Kyril Fitzlyon and Max Hayward (Bantam Books, New York, 1978), 228.
9 Abram Tertz, *Goodnight!*, trans. Richard Lourie (Viking Penguin, New York, 1989), 216.

1 THE PROVINCE OF PETER PORTER

1 Herman Melville, *Moby Dick* (Norton, New York, 1967), 56.
2 Seamus Heaney, *The Place of Writing* (Scholars Press, Atlanta, 1989).
3 Samuel Beckett, *All that Fall*, in *The Complete Dramatic Works* (Faber and Faber, London, 1986), 183.
4 Robert Lowell, 'Waking Early Sunday Morning', *Selected Poems* revised edition (Farrar, Straus and Giroux, New York, 1977), 141.
5 Karl Shapiro, 'The Death of Randall Jarrell', in *Randall Jarrell, 1914–1965*, eds Robert Lowell, Peter Taylor and Robert Penn Warren, (Farrar, Straus and Giroux, New York, 1967), 201.
6 W. H. Auden, 'The Sea and the Mirror', *Collected Poems*, ed. Edward Mendelson (Faber and Faber, London, 1976), 328.

2 THICKENING THE POETRY

1 Yvor Winters, 'At the San Francisco Airport', *The Hudson Review Anthology*, ed. Frederick Morgan (Vintage Books, New York, 1961), 215.
2 Peter Porter, Canberra mss.
3 W. H. Auden, *Collected Poems*, 414.
4 William Carlos Williams, quoted in Peter Conn, *Literature in America: An Illustrated History* (Cambridge University Press, Cambridge, 1989), 374.
5 Robert Frost, 'I would have written of me on my stone:/ I had a lover's quarrel with the world.' From 'The Lesson for Today', *Selected Poems* (Penguin Books, Ringwood, 1955), 231.
6 cf. Porter, 'Brisbane Comes Back', *Quadrant*, 98, XIX, 6 (September 1975), 52–8; and 'Locked Out of Paradise', *The New Review*, 36, (3 March 1977), 15–20.
7 The poet was Vincent Buckley, in conversation.
8 Elizabeth Jennings, 'Visit to an Artist', *Collected Poems, 1953–1985* (Carcanet Press, Manchester, 1986), 59.
9 Jonathan Swift, *Journal to Stella*, ed. Harold Williams; two vols. (Blackwell, Oxford, 1974).
10 Auden, 'September 1, 1939', *Selected Poems* new edition, ed. Edward Mendelson (Vintage Books, New York, 1979), 88.
11 Charles Lamb to William Wordsworth, 26 August 1816: '[Coleridge] is at present under the medical care of a Mr Gilman (Killman?) a Highgate Apothecary, where he plays at leaving off Laudanum. I think his essentials not touched: he is very bad, but then he wonderfully picks up another day, and his face when he repeats his verses hath its ancient glory, an Archangel a little damaged.' Quoted in Norman Fruman, *Coleridge, The Damaged Archangel* (Allen & Unwin, London, 1971), vii.
12 Walter Bagehot: 'There seems to be an unalterable contradiction between the human mind and its employments. How can a *soul* be a merchant? What relation to an immortal being have the price of linseed, the tare on tallow, or the brokerage on hemp? Can an undying creature debit *petty expenses* and charge for *carriage paid*? The soul ties its shoes; the mind washes its hands in a basin. All is incongruous.' Quoted in *The Faber Book of Aphorisms*, eds W. H. Auden and Louis Kronenberger (Faber and Faber, London, 1970), 9.
13 W. H. Auden, 'For the Time Being', *Collected Poems*, 279.
14 Jean Starobinski, *Montaigne in Motion*, trans. Arthur Goldhammer (University of Chicago Press, Chicago, 1985).
15 cf. Auden, 'Thanksgiving for a Habitat', in *Collected Poems*, 518–38.

16 cf. David Jones, *The Anathemata* (Faber and Faber, London, 1952).
17 Kenneth Burke, *Language as Symbolic Action: Essays on Life, Literature, and Method* (University of California Press, Berkeley, 1968).
18 cf. *The Shape of Poetry and the Shape of Music* for example (University of Tasmania Occasional Paper no. 28, 1980).
19 John Hollander, *Melodious Guile: Fictive Pattern in Poetic Language* (Yale University Press, New Haven, 1988).

3 ROMANCE

1 Alfred North Whitehead, *The Aims of Education and Other Essays* (Ernest Benn, London, 1932), 27–30.
2 ibid. 57.
3 Interview with Joseph Brodsky, *Writers at Work: The Paris Review Interviews: 8th Series*, ed. George Plimpton (Penguin Books, Ringwood, 1988) 385.
4 Seamus Heaney, 'The Diviner', *Selected Poems 1965–1975* (Faber and Faber, London, 1980), 24.
5 Howard Nemerov, 'Lion and Honeycomb', *The Collected Poems of Howard Nemerov* (University of Chicago Press, Chicago), 1977.
6 W. H. Auden, 'Making, Knowing, and Judging', *The Dyer's Hand and Other Essays* (Vintage Books, New York, 1968), 52.
7 Richard Wilbur, 'Junk', *New and Collected Poems* (Harcourt Brace Jovanovich, New York, 1988), 185.
8 Marianne Moore, 'Poetry', *The Complete Poems of Marianne Moore* (Penguin Books, Ringwood, 1982), 36.
9 Porter, 'On First Looking into Chapman's Hesiod', *Collected Poems*, (Oxford University Press, Oxford, 1983), 210: 'Country Poetry and Town Poetry: A Debate with Les Murray', *Australian Literary Studies*, 9, 1 (1979), 39–48.
10 W. H. Auden, *The Dyer's Hand*, 354.
11 Michael Frayn, *Constructions* (Wildwood House, London, 1974). Paragraphs, not pages, are numbered.
12 Wallace Stevens, 'Selections from "Adagia"', *Modern Poets on Modern Poetry*, ed. James Scully (Collins Fontana, London, 1966), 157.
13 W. S. Merwin, 'Home for Thanksgiving', *Selected Poems* (Atheneum, New York, 1988), 81.
14 A. D. Hope, 'The Wandering Islands', *Collected Poems 1930–1970* (Angus & Robertson, Sydney, 1972), 26.
15 Robert Browning, 'Andrea del Sarto', *The Poems of Robert Browning*, vol. I, ed. John Pettigrew (Penguin Books, Ringwood, 1981), 643.

16 F. R. Leavis, 'The Irony of Swift', *Scrutiny*, ii (March 1934), 364–78.
17 In *The Oxford Book of Prayer*, ed. George Appleton (Oxford University Press, Oxford, 1985), 297.
18 For example, in 'Horae Canonicae', *Collected Poems*, 475; and in 'Reading', *The Dyer's Hand*, 6, 7.
19 W. H. Auden, 'In Praise of Limestone', *Collected Poems*, 414.
20 Emily Dickinson, 'A narrow Fellow in the Grass', *Complete Poems*, ed. Thomas H. Johnson (Faber and Faber, London, 1975), 459.
21 *The Pocket Book of Ogden Nash* (Washington Square Press, New York, 1962), 171.

4 PRECISION

1 Michael Meszaros. The dictum is on a brass plaque on the Surveying and Land Information Building, at the University of Melbourne.
2 For reflections on the 'tombeau' modes of writing, see Lawrence Lipking, *The Life of the Poet: Beginning and Ending Poetic Careers* (University of Chicago Press, Chicago, 1981), 138–60.
3 T. S. Eliot, 'Little Gidding', *The Norton Anthology of Modern Poetry*, eds Richard Ellmann and Robert O'Clair (Norton, New York, 1973), 473.
4 Richard Wilbur, 'Tywater', *New and Collected Poems*, 342.
5 Porter, 'The Recording Angels: Music and Meaning in the Voices of Poetry', *Times Literary Supplement*, 4589, (15 March 1991), 3–4.
6 John Donne, 'The Canonization', *The Complete Poetry and Selected Prose of John Donne*, ed. Charles Coffin (Modern Library, New York, 1952), 14.
7 The Gospel According to Matthew, XXV, 23.
8 'Little Harmonic Labyrinths: An Interview with Peter Porter', *Southerly*, 2 (1985), 12–22.
9 The answers here are: middle ('Talking to the Lizards', *PW*, 59); middle ('Civilization and Its Disney Contents', *PW*, 35); beginning (The Wind at Bundanon', *PW*, 12); end ('At Schubert's Grave', *CB*, 33) .
10 W. H. Auden, *Collected Poems*, 608.
11 Blaise Pascal, *Pensées*, ed. Louis Lafuma, trans. John Warrington (Dent, London, 1973), 264–5.
12 Peter Porter, 'Barding it up in Bunyah', *Scripsi*, 5, 1 (1988), 195.

5 GENERALISATION

1 Peter Porter, 'Grub Street Versus Academe', *The New Review*, 1, 4 (1974), 52.
2 Peter Porter, 'The Shape of Poetry and the Shape of Music', *CP* 7, 8.

3 Georg Christoph Lichtenberg, *Aphorisms*, trans. R. J. Hollindale (Penguin Books, Ringwood, 1990), 30, 36, 44, 42, 82.
4 Samuel Johnson, 'The Vanity of Human Wishes', *The Norton Anthology of Poetry* revised edition, eds Alexander W. Allison et al., (Norton, New York, 1970), 496.
5 Peter Porter, 'Interview with Peter Porter', *Westerly*, 35, 2, 62.
6 Aristotle, *Rhetoric and Poetics*, trans. W. Rhys Roberts (New York, Random House, 1954). Quoted in Warren A. Shibles, *Metaphor: An Annotated Bibliography and History* (Language Press, Whitewater Wisconsin, 1971), 31.
7 Peter Porter, 'Little Harmonic Labyrinths', *CP* 22.
John Keats, 'Ode on a Grecian Urn', *Romantic Poets: Blake to Poe*, ed. W. H. Auden and Norman Holmes Pearson (Viking, New York, 1950), 385.
9 John Keats to Bailey, November 22, 1817: 'The Imagination may be compared to Adam's dream—he awoke and found it truth.' Quoted in Ben Belitt, *Adam's Dream: A Preface to Translation* (Grove Press, New York, 1978), 5.
10 Percy Bysshe Shelley, in *Shelley's Critical Prose*, ed. Bruce R. McElderry, Jnr (University of Nebraska Press, Lincoln, 1967), 6.
11 The saying is quoted by Yeats from Coventry Patmore, and is cited by Heaney in, among other places, *Preoccupations: Selected Prose 1968–1978* (Faber and Faber, London, 1980), 7, and 'The Harvest Bow', in Heaney, *Field Work* (Faber and Faber, London, 1979), 58.
12 Peter Porter, in conversation, in Melbourne, 1991.
13 Nicolas Malebranche, quoted in Henry de Lubac, *The Discovery of God*, trans. Alexander Dru (Darton, Longman and Todd, London, 1960), 52.
14 Georg Christoph Lichtenberg, *Aphorisms*, 79.
15 Seamus Heaney, 'Making Strange', in *Station Island* (Faber and Faber, London, 1984), 32.
16 W. H. Auden, 'In Praise of Limestone', *Collected Poems*, 414.
17 Franz Kafka, *The Great Wall of China: Stories and Reflections*, trans. Willa and Edwin Muir (Schocken Books, New York, 1970), 163, 170.
18 John Olsen, 'Foreword', in Robert Hughes, *Donald Friend* (Edwards and Shaw, Sydney, 1965), 7.
19 Ludwig Wittgenstein, quoted in Karsten Harries, 'The Many Uses of Metaphor', *On Metaphor*, ed. Sheldon Sacks (University of Chicago Press, Chicago, 1979), 167.
20 Peter Porter, in 'Saving from the Wreck', *Scripsi* 3, 2 and 3 (1985), 196.
21 Peter Porter, 'The Shape of Poetry and the Shape of Music', *CP* 2.

22 Chamfort, *Products of the Perfected Civilization*, trans. W. S. Merwin (North Point Press, San Francisco, 1984) 120, 123, 179, 251.
23 Jean Seznec, *The Survival of the Pagan Gods: The Mythological Tradition and Its Place in Renaissance Humanism and Art*, trans. Barbara F. Sessions (Princeton University Press, New York, 1953).
24 Leonard Barkan, *The Gods Made Flesh: Metamorphosis and the Pursuit of Paganism* (Yale University Press, New Haven, 1986).
25 John Dryden, 'Absalom and Achitophel', in *Selected Poetry and Prose of John Dryden*, ed. Earl Miner (Modern Library, New York, 1969) 206.
26 W. S. Merwin, *The Miner's Pale Children* (Atheneum, New York, 1981), 163, 174.
27 Leonard Barkan, *The Gods Made Flesh*, xv.
28 Eva Hoffman, *Lost in Translation: Life in a New Language* (Minerva Books, London, 1989).
29 Peter Porter, 'Locked out of Paradise', *CP* 15, 16.
30 Peter Porter, 'Saving from the Wreck', *CP* 199.
31 Peter Porter, 'Miracles of Metamorphosis: Gustav Mahler', *Encounter*, (August 1976), 56, 60.
32 Peter Porter, 'Auden's Cornucopia: The 1930s Texts', *Encounter*, (February 1978), 70.
33 A. D. Nuttall, *The Stoic in Love: Selected Essays in Literature and Ideas* (Harvester Wheatsheaf, London, 1989), 188.
34 Lichtenberg, *Aphorisms*, 56.

6 THREE POEMS—SOME POINTERS

1 Rainer Maria Rilke, *Selected Poems*, ed. and trans. Stephen Mitchell (Picador, London, 1987), 59.
2 Robert Browning, 'Balaustion's Adventure', *The Poems of Robert Browning*, vol. 1, 867–942.
3 John Pettigrew, *The Poems of Robert Browning*, vol. 1, 1170.
4 ibid. 'Balaustion's Adventure', 940.
5 ibid. 936.
6 John Donne, 'Satyre III', *Complete Poetry and Selected Prose*, 97.
7 Gavin Ewart, *The Collected Ewart, 1933–1980* (Hutchinson, Sydney, 1980), 355.
8 Peter Porter, 'Little Harmonic Labyrinths', *CP* 17–18.
9 Ralph Waldo Emerson, quoted in *The Faber Book of Aphorisms*, eds. W. H. Auden and Louis Kronenberger (Faber and Faber, London, 1970), 275.

10 Wallace Stevens, in *Modern Poets on Modern Poetry*, 155.
11 Robert Lowell, *Poems 1938–1949* (Faber and Faber, London, 1950), 79.
12 Ibid. 55.
13 William Shakespeare, *King Lear*, IV, i, 36.
14 John Pearson, *Arena: The Story of the Colosseum* (Thames and Hudson, London, 1973).
15 Pearson, *Arena*, 39, 41, 46, 52–3, 153, 155, 187.
16 Charles Péguy, *Temporal and Eternal*, trans. Alexander Dru (Harvill Press, London, 1958), 48.
17 W. B. Yeats, 'Per Amica Silentia Lunae', *Mythologies* (Macmillan, London, 1962), 331.
18 John Donne, 'A Valediction: Forbidding Mourning', *Complete Poetry and Selected Prose*, 38.
19 Marianne Moore, 'Spenser's Ireland', *Complete Poems*, 112.
20 Randall Jarrell, 'Deutsch Durch Freud', *The Complete Poems*, 266.
21 Pearson, *Arena*, 176.
22 Marguerite Yourcenar, *Memoires d'Hadrien* (Gallimand, Paris, 1988), 307. (Translation is by Christopher Willcock, SJ)
23 W. B. Yeats, 'The Second Coming', *Collected Poems*, 187.
24 Psalm 137.
25 Plutarch, *The Life of Cicero*, trans. J. L. Moles (Aris and Phillips, Warminster, 1988), 143.
26 G. K. Chesterton, 'The Ballad of the White Horse', *The Collected Poems of G. K. Chesterton* (Methuen, London, 1942), 265.
27 W. H. Auden, *Collected Poems*, 444.
28 Thomas Wyatt, 'Mine Own John Poins', *The Norton Anthology of Poetry*, 121–2.
29 W. B. Yeats, *Autobiographies*, (Macmillan, London, 1955), 527.

BIBLIOGRAPHY

PRINCIPAL WORKS BY PETER PORTER

John R. Kaiser's *Peter Porter: A Bibliography 1954–1986* (Mansell Publishing, London and New York, 1990) is a comprehensive guide to Porter's publications, in prose and in verse, up to the mid-eighties.

Manuscripts and Notebooks

Most of the notebooks and looseleaf manuscripts containing versions of Porter's poetry and some prose are held at the Australian National Library, Canberra. Others are to be found at the Lockwood Memorial Library, Buffalo; at the British Museum; at Indiana University; and at Reading University Library.

Books of Poetry

Once Bitten, Twice Bitten (Scorpion Press, Northwood, Middlesex, 1961).
Poems Ancient and Modern (Scorpion Press, Lowestoft, Suffolk, 1964).
A Porter Folio (Scorpion Press, Lowestoft, Suffolk, 1969).
The Last of England (Oxford University Press, London, 1970).
After Martial (Oxford University Press, London, 1972).
Preaching to the Converted (Oxford University Press, London, 1972).
Jonah by Arthur Boyd and Peter Porter (Martin Secker and Warburg, London, 1973).
The Lady and the Unicorn by Arthur Boyd and Peter Porter (Martin Secker and Warburg, London, 1975).
Living in a Calm Country (Oxford University Press, London, 1975).
The Cost of Seriousness (Oxford University Press, Oxford, 1978).
English Subtitles (Oxford University Press, Oxford, 1981).
Collected Poems (Oxford University Press, Oxford, 1983).
Fast Forward (Oxford University Press, Oxford, 1984).
Narcissus by Arthur Boyd and Peter Porter (Martin Secker and Warburg, London, 1984).
The Automatic Oracle (Oxford University Press, Oxford, 1987).
Mars by Arthur Boyd and Peter Porter (André Deutsch, London, 1988).
Possible Worlds (Oxford University Press, Oxford, 1989).
A Porter Selected (Oxford University Press, Oxford, 1989).
The Chair of Babel (Oxford University Press, Oxford, 1992).

Interviews

Douglas, Denis, 'Conversation with Peter Porter', *Overland*, 44 (Winter 1970), 33–4.

'The Poet in the Sixties: Vices and Virtues—a recorded conversation with Peter Porter', in Michael Schmidt and Grevel Lindop, *British Poetry Since 1960: A Critical Survey* (Carcanet Press, South Hinksey, 1972), 202–12.

Martin, Philip, 'Interview with Peter Porter', *Quadrant*, 87, XVIII, No. 1 (Jan–Feb. 1974), 9–19.

Bennett, Bruce, 'Peter Porter in Profile', *Westerly*, 27, 1 (March 1982), 45–56.

Harrison, Martin, 'Peter Porter. Interviewed by Martin Harrison', *Australian Literary Studies*, 11, 4 (October 1984), 458–67.

Kavanagh, Paul, 'Little Harmonic Labyrinths: An Interview with Peter Porter', *Southerly*, 1 (March 1985), 12–22.

Peter Porter: What I Have Written, Writers Talking (Australian Film Institute video, 1988); 30 minutes.

Baker, Candida, *Yacker 3. Australian Writers Talk About Their Work* (Pan Books, Sydney, 1989), 260–83.

Williams, Barbara, 'An Interview with Peter Porter', *Westerly*, 35, 2 (June 1990), 57–73.

Articles and Commentaries

'Grub Street Versus Academe', *The New Review*, 1, 4, (1974), 47–52.

'Byron and the Moral North', *Encounter* (August 1974), 65–73.

'Return to Arcadian Australia', the *Listener*, 92, 2372 (1975), 333.

'Miracles of Metamorphosis: Gustav Mahler', *Encounter* (August 1976), 55–60.

'Brisbane Comes Back', *Quadrant*, 98, XIX, 6 (September 1975), 52–8.

'Locked Out of Paradise', *The New Review*, 36, 3 (March 1977), 15–20.

'Country Poetry and Town Poetry: A Debate with Les Murray', *Australian Literary Studies*, 9, 1 (1979), 39–48.

'The Shape of Music and the Shape of Poetry', *Quadrant* XXXIV, 6 (June 1980), 4–12.

'The Professional Amateur', *Spirit of Wit: Reconsiderations of Rochester*, ed. Jeremy Treglown, Blackwell, Oxford, 1982, 58–74.

'Melting Moments: Puccini and his Critics', *Encounter*, LIX, 5 (November 1982), 46–52.

'The Mystery and the Music: On Mozart's Life', *Encounter*, LX, 6 (June 1983), 53–8.

'Saving from the Wreck', *Scripsi* 3 2 and 3 (1985), 195–216.

'Working with Arthur Boyd', *Westerly*, 32, 1 (March 1987), 69–78.

'An Expatriate's Reaction to his Condition', *Westerly*, 32, 2 (December 1987), 43–7.

'Barding it up in Bunyah', *Scripsi*, 5 1 (1988), 191–6.

'Incandescence of the Ordinary', *Overland*, 113 (December 1988), 14–16.

'The Recording Angels: Music and Meaning in the Voices of Poetry', the *Times Literary Supplement*, 4589 (15 March 1991), 3–4.

PRINCIPAL REFERENCES ABOUT PETER PORTER

Books

Bennett, Bruce, *Spirit in Exile: Peter Porter and His Poetry*, Oxford University Press, Melbourne, 1991.

Articles and Commentaries

Bennett, Bruce, 'Passports to Paradise: Peter Porter and Clive James', *A Sense of Exile: Essays in the Literature of the Asia–Pacific Region*, eds Bruce Bennett and Susan Miller (The Centre for Studies in Australian Literature, University of Western Australia, Nedlands, 1988), 67–80.

Brownjohn, Alan, 'From the Eighth Floor of the Tower: The Collected Peter Porter', *Encounter* 358 (Sept.–Oct. 1983), 80–4.

Carter, Paul, 'An Effect of Distance: Porter's History', *Overland*, 117 (Feb. 1990), 20–4.

Ewart, Gavin, 'From Notions to Emotions', the *Times Literary Supplement* (19 May 1978), 550.

Garfitt, Roger, 'The Group', *British Poetry Since 1960: A Critical Survey*, eds M. Schmidt and G. Lindop (Carcanet Press, South Hinkset, Oxford, 1972), 15–59.

Hulse, Michael, 'Love and Death: Nine Points of Peter Porter', *Quadrant*, 193, XXVII, 9 (September 1983), 31–8.

James, Clive, 'Settling for Dust', *The Metropolitan Critic* (Faber and Faber, London 1974), 66–75.

James, Clive, 'The Boy from Brisbane', *Snakecharmers in Texas: Essays 1980–87* (Picador, London, 1989), 44–8.

Jones, Evan, 'The Poetry of Peter Porter', *Scripsi* 2 4 (1984), 65–75.

Morrison, Blake, 'A Philosopher of Captions', the *Times Literary Supplement*, 4174 (1 April 1983), 321–2.

Murray, Les, 'On Sitting Back and Thinking About Porter's Boeotia', in *The Peasant Mandarin: Prose Pieces by Les Murray* (University of Queensland Press, St Lucia, 1978), 172–84.

Pollnitz, Christopher, 'Peter Porter: Whether "The World is But a Word"', *Scripsi* 5 1 (1988), 197–208.

Rawson, Claude, 'Moving Pictures', *London Review of Books* (16 July–5 Aug. 1981), 14–15.

Steele, Peter, 'The Radiations of Peter Porter'. *Westerly*, 29 3 (October 1984), 65–74.

Wallace-Crabbe, Chris, 'The Ingredients Are Myself', *Top Shelf Australian Book Review* (1991), 26.

ACKNOWLEDGEMENTS

I have left Peter Porter untroubled by enquiries about his poetry, but a friendship of some years' standing yields letters and conversations: for these, as for the poetry, my gratitude matches my pleasure. Chris Wallace-Crabbe invited me to write for Oxford's 'Australian Writers' series, and has kept a comradely eye on what has emerged. Peter Rose, my editor at Oxford, has combined geniality with acuity in a very welcome way. Bruce Bennett kindly let me read in manuscript his *Spirit in Exile: Peter Porter and His Poetry*, to my great profit. Librarians at the National Library of Australia were helpful. My Jesuit confrères at Newman College, Melbourne, have been more encouraging than they know. A book must take its own chances, but there is nothing chance about my thanks to all these people.